Global Warming Temperatures and Projections

As Related to CO_2 and H_2O Absorptions, H_2O Evaporation, and Post-Condensation Convection

Unbiased Mathematical Calculations
and Calibration Plots of the Effects
of Carbon Dioxide and Water

William T. Lynch, PhD

Lake Villa District Library
Lake Villa, Illinois 60046
(847) 356-7711

Copyright © 2017 William T. Lynch PhD
All rights reserved.
ISBN 978-1-365-92746-1

All rights reserved. This book or any portion thereof may not be reproduced or used in any manner whatsoever without the express written permission of the publisher except for the use of brief quotations in a book review or scholarly journal.

First Printing: 2017

Discounts are available on quantity purchases. Contact the author at globalwarming.lynch@gmail.com.

Contents

Preface 1
Abstract (extended) 3
Preliminary 'peeks' 21

Section One: Modeling the atmosphere
   Introduction 27
   Definitions 31
   A very brief summary of the spreadsheet model 37
   Normalization of altitude 39
   Basic BBR, power and Planck expressions 41
   Calculations within each slice 43
   A second representation of the notch 51
   The overall picture of atmospheric interactions 53
   Molecular density and altitude 59
   Soft entry to the calibration tables 61
   Calibration 63
   Continuation of the (1-TOth) & (1-Tcomb) analysis 73
   Doubling plots: $\Delta K(0)$ vs. $LOG2(CO_2/380)$ 77
   Comparisons with the calibration plot 85
   Additional details on absorption 91
   How much additional absorption does CO_2 provide? 97
   Progression of atmospheric parameters 101
   Section One summary 107
   Section One references 111

Section Two H_2O and its phases
   H_2O water vapor vis a vis CO_2 115
   The adiabatic lapse rate 119
   Specific comparisons of atmospheric water and CO_2 .. 123
   Average water vapor content in the atmosphere 127
   Published distributions of water vapor 129
   Heat convection from the Tropics 139
   Water as a 'climate changer 151
   Section Two summary 157
   Section Two references 161

Appendicies
  Appendix A: Correlations with total absorption 165
  Appendix B: Normalized altitude Z' 169
  Appendix C: Organization of the spreadsheet 175
  Appendix D: Absorption 'notches' 189
  Appendix E: Final sets of data and their interpretation ... 201
Quick references (5 tables)............................ 221
About the author 227

Acknowledgments

William T. Lynch acknowledges the early guidance and comments by Professors Hugo Fritz Franzen (Iowa State) and Will Happer (Princeton). He also acknowledges the book compilation and editing by Anthony Lynch, the helpful commentaries by Kevin Johnson, Edward Stubenrauch and Joseph Kneuer, and the continuing support of this project by his wife, Gerry Lynch.

Global Warming Temperatures and Projections 1

Preface

The analyses of this document were initiated neither as a polemic against nor an apologia for existing opinions about CO_2 and its role in global warming.

Its primary purpose is to quantify convincingly the limiting effects of increases in CO_2 on the earth's temperature and to do so independently of any possible linkages to secondary enhancements. In particular, any propensities for CO_2 to produce runaway effects had to be resolved.

A secondary purpose is to assess why official projections for global warming have been in disagreement with the official measurements, and to produce a model that can adjust parameters so as to, forcibly if necessary, cause official CO_2 and temperature values to be in agreement.

There has been no verification that official average (as opposed to peak) daily temperatures have been uniformly and properly calculated and integrated over many years of increasing data points with new technologies and new locations. Official records of past averages have been accepted, but official projections have not been employed.

Preface

A truly successful goal would be a calibration plot that aligns average surface temperature K(0) with whatever absorbing molecules are present in the atmosphere. Any additions in surface temperatures should more likely be self-limiting rather than runaway scenarios.

These goals have been achieved, but all CO_2 and H_2O increases in absorption are, by themselves, insufficient to provide the necessary increases in the Blanket to support the K(0) values of the calibration plots.

Section One incorporates all the arguments associated with CO_2 and H_2O absorption.

Section Two incorporates the additional, and significant, effects associated with evapotranspiration, surface cooling, upper atmosphere condensation, and the lateral convection of the heats of condensation from the Tropics latitudes towards the Poles.

This convection provides almost half the Blanket in the Temperate zones, reduces temperature in the Tropics, and has little to nothing to do with CO_2 or other human influences.

The calibration plots of Section One are made whole again, without absolute proof, by the realization that the unspecified "background" absorptions are directly replaceable with the justifiable convection contributions to the blanket.

The document is organized as a sequence of learning elements. The ultimate conclusions are firmly based in physics and mathematical constructs.

The openness of the analysis has been rewarded by proofs that the effects of CO_2 have been overstated, that increases in global warming are limited, and that nature does not support a runaway.

The style of this book intentionally avoids having hundreds of references. References with direct application are acknowledged. References are not sought for knowledge that is believed to be well-accepted.

Abstract (extended)

Section One concentrates on CO_2 and on physics- and mathematics-based models. The emphasis is on calculating absorptions by CO_2 molecules of Black Body Radiation (BBR) from the surface and in the return of nearly one half of that absorbed energy in the form of a blanket. Section Two emphasizes H_2O and its phases. H_2O not only has an atmospheric molecular phase (vapor) but also an atmospheric liquid phase, which is, of course, critical for rainfalls. Section Two also emphasizes the crown of atmospheric heat in the Tropics that spills towards the Poles and contributes significantly to local blankets.

Section One presents an unbiased mathematical confirmation of a general model that states that any absorbing molecules in the atmosphere will interrupt surface heat from escaping directly. The returning "blanket" raises the surface temperature K(0) and the surface BBR emission until a level is reached at which the unabsorbed portion plus the half that is not returned as a blanket is sufficient to

Abstract (extended)

satisfy the demand for K(0) equilibrium. In all of these calculations thermodynamic adiabatic effects must be considered. Energies are never lost, but adiabatic work is always taking place. (See illustration at right.)

Absorptions by present CO_2 levels are far from sufficient to match the levels of the actual BBR, and Section One determines the level of additional absorptions that must be present in order to match present reality. H_2O concentrations are also introduced in Section One; direct H_2O absorptions are equivalent to CO_2 but the combined CO_2 and H_2O absorptions are still insufficient to satisfy present reality. There is a residual "background" of apparent absorption that has to be explained. (See **Figure A1**.)

> For a guide to CO_2 and energy in the atmosphere and oceans, of natural and human origin, refer to
> **Quick references (5 tables) following the appendicies**

Section One deals only with a vertical modeling of the atmosphere, with an average surface temperature K(0). All net directions are up and down, because the model block has an undefined surface area that extends to left and right and encompasses the world. Its ultimate conclusions are, in fact, accurate, except that the residual background tentatively incorporated as an unknown absorption does not have to be totally ascribed to absorptions. Section Two shows that the "unknown absorption" is primarily a blanket supplement dominated by upper atmosphere heat convection emanating from the Tropics. (See **Figures A2, A3, A4**.)

Although both H_2O and CO_2 are necessary inputs for plant growth, CO_2 never appears in the liquid or solid phases, and so does not make its presence in the atmosphere known by means of evaporation, but H_2O does. Furthermore, the evaporation of water from the surface significantly *cools* the surface. The required heat of evaporation of water is very high; it takes more than 500 times as much heat to vaporize one gram of water as to raise its liquid

Global Warming Temperatures and Projections

Cross section parameters for the earth and atmosphere

Z = Zcutoff ≅ 13km

P(Z=13 km) is approx pinned at 97W/m²

Negligible adiabatic effects for Z > Zcutoff ≅ 13km

Temperature K, molecular density, and thermalized energy (and power) decrease with altitude.

There is an adiabatic reduction of any upwards heat flow because of the thermodynamic "work" associated with the reduction of volume density

Z = 0 (earth's surface)

P(0) = BBR(0) ≅ 390 W/m²
Ave K(0) = 288K (15C, 59F)
Local K(0)'s = 288 ± 40K

Earth and atmosphere values

Earth
- Area ≅ 512E12 m²
- Radius R ≅ 6,370 km
- Zcutoff/R ≅ 0.0021
- Average K(0) ≅ 288K
- K(0)'s may vary by ± 40K
- Ave P(0) = BBR(0) ≅ 390 W/m²

Atmosphere
- Total # of molecules ≅ 1E44
- Molecules/m² ≅ 1.96E29/m²
- Molecules below Z = 13 km ≅ 1.6E29/m²

temperature by 1.0C. *Cooling* takes place upon evaporation of liquid water but *heating occurs higher in the atmosphere when the vapor is condensed.*

Besides any long-term increase in atmospheric CO_2 from carbon burning, atmospheric CO_2 does demonstrate cyclic shifts when temperatures change from any cause (such as El Niño and La Niña

Abstract (extended)

Figure A1: Basic model with no evaporation

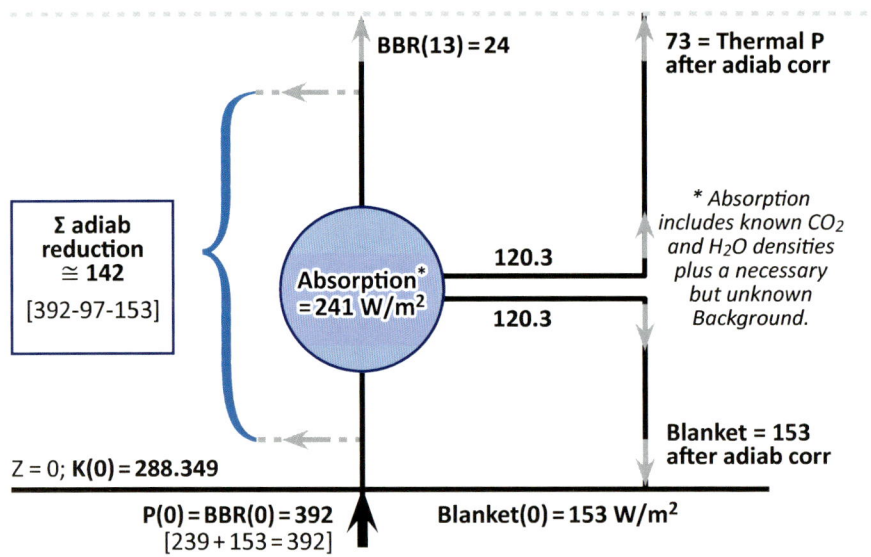

In this example, CO₂=632 ppm, PWV=5 cm, and K(0)=288.349K. This result is unrealistic; it requires additional "background" absorbers with absorption capabilities much greater than for the combined CO₂ and water vapor. Tbkgnd is 0.2758.

effects). The oceans are a large reservoir of CO_2 and release CO_2 when warmed and collect CO_2 when cooled. (The daily net releases and collections can at times be greater than the human production of new CO_2.) CO_2 has no direct linkage to produce unusual weather effects but the multiple effects for creating a constantly varying H_2O atmospheric distribution — pressure, temperature, surface water, Coriolis effect, convection — absolutely disrupt weather uniformity. Nature does not adjust uniformly and temporally towards a global average, but adjusts locally and instantaneously to disruptions caused by sun changes, wind changes, and the 23 degree offset of the earth's rotational axis relative to the earth's axis of rotation

Figure A2: *The cooling effect of water evaporation*

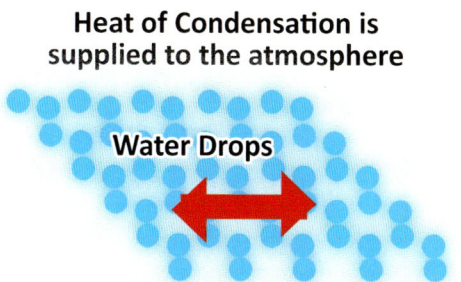

Heat of Condensation is supplied to the atmosphere

Water Drops

There may be some convection of heat along this channel, or some radiation to the earth, or some (warmed) water may fall to earth

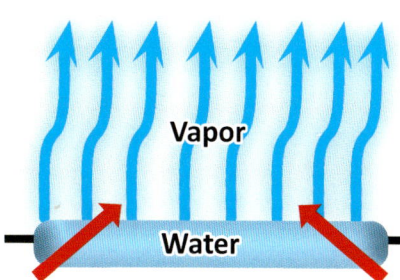

Vapor

Water

Heat of Vaporization cools the earth

An average rainfall of 2 mm/day (29 inches per year) requires a restoration of 2 mm/day. This amount of evaporation corresponds to a cooling of 52 W/m^2. With no return of this heat, the ΔK(0) would be about -10.0K (-10.0°C)

---

around the sun. The offset does more than just cause seasonal opposites between the northern and southern hemispheres; for latitudes beyond the range of the 46 degree Tropical band the winters are colder and the summers are warmer than they otherwise would be.

All in all, this may be considered as a fortunate result, but it produces more weather disruption at all latitudes. Weather disruption is greatest at latitudes where the earth's rotational speeds are large and, simultaneously, where the rotational speeds vary significantly with latitude. The mid-latitudes, therefore, are the optimum latitudes for hurricanes and tornados. (They are optimum for nature, not necessarily so for humans.) Nature's power and energy

8 Abstract (extended)

Figure A3: Without local evaporation and conv = 80 W/m²

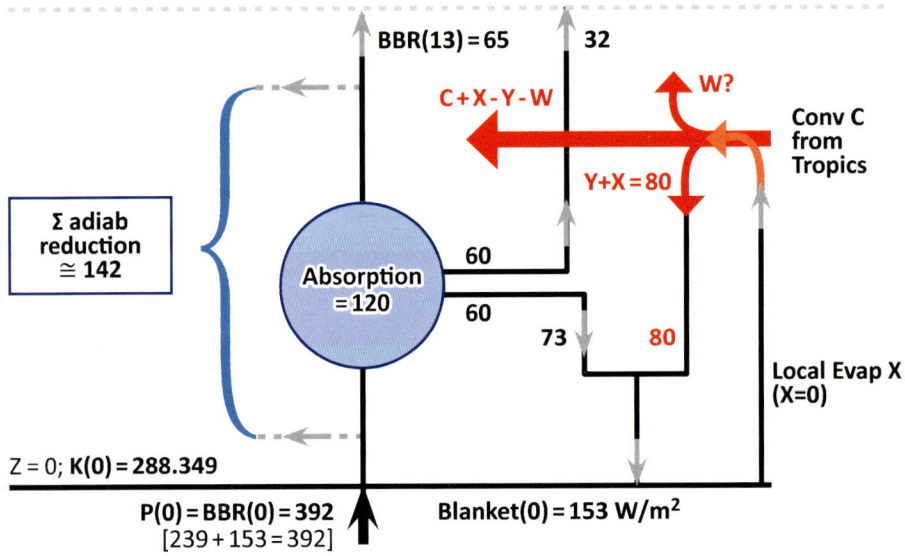

CO₂= 632 ppm; PWV = 5 cm; **Local Evap = 0; Additional Conv Blkt = 80;** Z = 13 km; BBR(13) = 65 W/m²; Thermal P ≅ 32; Total P ≅ 97; K ≅ 203.5K

In this example, *eff*TCO₂=0.7336; *eff*TH₂O (PWV)=0.7760; *eff*TH₂O (con-vBlkt)=0.2974; Tbkgnd=0.9272. Tcomb=0.1570 and (1-Tcomb)=0.8430, which is exactly what it should be for K(0)=288.349K. The additional blanket dominates over the combined CO₂+H₂O molecular absorption. Any remaining background absorptions (CH₄, etc.) are contained within Tbkgnd.

contributions to these local events are phenomenally larger than total worldwide power and energy creation by humans. (See **Quick references (five tables)** after the appendicies.)

Circulation cells

Section Two contains explanations of the three globe-circling tubes (known as "circulation cells") within each hemisphere (**Figure 43**). The equator is a "Low" pressure with heavy rains. Tropical heat is associated with hot air winds projected from about 24°L toward the

Global Warming Temperatures and Projections

Figure A4: *Atmospheric model with convection bus*

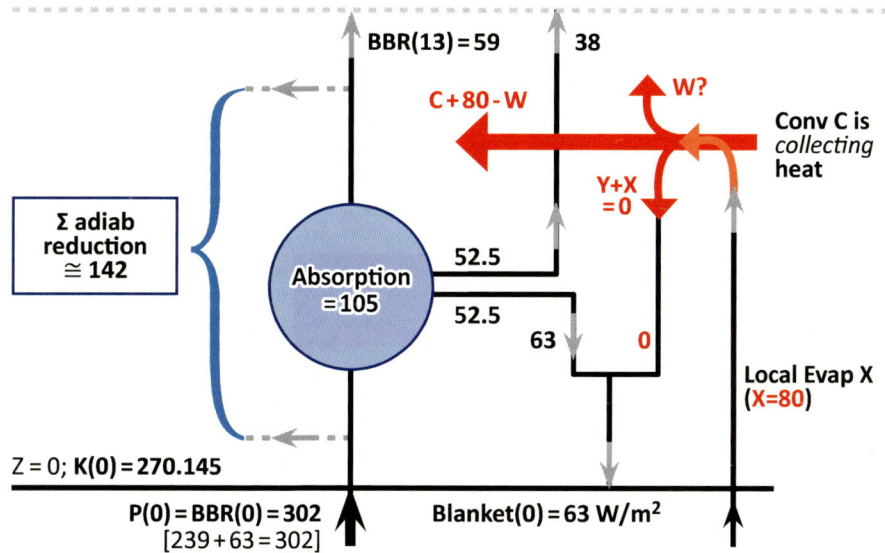

CO₂= 380 ppm; PWV = 5 cm; **Local Evap = 80; Additional Conv Blkt = 0;** Z = 13 km; BBR(13) = 59 W/m²; Thermal P ≅ 38; Total P ≅ 97; K ≅ 203.5K

This example is a dramatic flip from the previous example. The convection bus is collecting heat from the condensation of (intense) local evaporations but is not adding (returning) any additional blanket. There is a direct cooling effect on K(0), one that mimics the effect in the high temperature Tropics. Absorptions and BBR(13) are changed only slightly, but there is also a dramatic drop in the required value for Blkt(0). Tbkgnd is 0.821.

Poles and initially at relatively high altitudes. This heat begins to sink at about 30°L creating a "High." At the same time Polar cold winds are being projected towards lower Latitudes, initially at relatively low altitudes and the gradual heating produces a rise (and a "Low") at about 60°L. Looking from West to East, the rotations of the 0° to 30°L ("Hadley") and 60° to 90°L ("Polar") cells are counterclockwise (in the northern hemisphere). Nature introduces an intermediate coupling ("Ferrel") cell with a clockwise rotation that eliminates the direct up-down crashes that would take place if there were only two rotational cells. The mixing of the moisture-bearing winds from the Polar and Ferrel cells produces considerable, and annually variable, precipitation (at the "Low") near 60°L (**Figure 44**). This is evident in one of the Figures in Section Two. Moisture is much lower in the downward-directed flows (at the "High") near 30°L, and 30°L is, in fact, a favored latitude for deserts. The flexing Ferrel cell reduces heavy clashes, but the variation in that supposed edge at 60°L is a prime mitigator to more extreme climate change. The "polar vortex" is nothing more than the pulsing of that 60°L boundary during the winter. Nature cannot be tamed, however, and humans demonstrate great hubris if they try to alter nature's climate changes. Examples are given of the energy in hurricanes relative to mankind's creation of electrical energy.

Much evaporation, and atmospheric vapor heating, also occurs from the direct absorption of photons from the sun. Water vapor will eventually cool at upper altitudes, but there is a crown of warm saturated air in the Tropics (**Figures 38, 39**). Just as a cool morning at earth level produces dew droplets, higher atmosphere water vapor initially created from the evaporation of water (or other forms of evapotranspiration) also condenses into droplets. This can all be processed by the mathematical mechanics and temperature-dependent physics of Section One, but only if local cooling evaporation at earth level were equal to the blanket contribution from the upper atmosphere. Evaporations from the surface in the Temperate zone may be lower than the blanket contribution from the Tropics and the additional blanket contribution raises $K(0)$ in the Temperate zone. The net effect within the Tropics is actually negative since the outflow towards the Poles reduces surface temperatures in the Tropics by 12°C. Averaged temperatures over 60 years show no significant changes (**Figure 37**).

Calibration plots

The Section One calculations, including a calibration plot of surface temperature $K(0)$ vs. effective absorption, as represented by the full blanket's equivalence to an absorption, is the supporting basis for all of the additional observations and tabular conclusions in Section

Global Warming Temperatures and Projections

Two (**Figures 13, 16**). In Section Two the residual "background" of Section One is comfortably and convincingly — but with limited mathematical rigor — correlated with the evaporation, condensation, and convection associated with H_2O's natural responses to both the temperature and the rotational properties of the earth. Of great significance is the fact that the Tropics surface area is about 40%, the Temperate surface areas are about 51%, and the Polar surface areas are about 9%. The Tropics dominate the temperatures for all Latitudes and the sun dominates the Tropics (**Figure 40**).

A slice model of the atmosphere, with molecular counts rather than altitude differences as slice "thicknesses," transforms altitude Z into a normalized (dimensionless) Z' that ranges from 0 to 1.0 (**Figure AB3**). With an averaged temperature for each slice, and with deliberate consideration of absorption and adiabatic loss for WN bands within the Planck spectrum for each slice, analyses of BBR (directed) and thermalized (isotropic) power have been successfully applied to the question of global warming. The calculations within each slice and the matched boundary conditions to adjacent slices provide the bulk of the calculations for determining absorptions and power flows (**Figure AC3**).

Carbon dioxide's basic properties as a function of wavenumber (inverse wavelength) are easily modeled. The carry-forward of the WN dependencies from slice to slice is crucial to the final accuracy, since "notches" are quickly established in the residual BBR spectrum (**Figure 7**). These notch developments with Z' have been plotted. (See **Appendix D**.) Higher CO_2 concentrations, with higher temperatures at all altitudes, do not noticeably affect the notch depths at any altitude for those WN values that have shown a 99% notch for a lower CO_2 value. Notches become deeper and wider for all other WN values.

Increases of CO_2 by 2X to 20X affect wavenumbers with low sensitivity and slowly increase their absorption depths and only slowly increase K(0). (See **Figure 3**.) CO_2 by itself presents no threat to global warming. CO_2 is, in fact sensitive to only about one half of the BBR power spectrum. Currently, the *eff*ective transmission coefficient for infrared BBR, based upon only CO_2, is $effTCO_2 \cong 0.75$, where 'effective' means averaging over all WN and including the effects of

"Other" competing absorbers. For very recent climate conditions (CO_2=380 ppm, K(0)=288K} the overall "combined" Tcomb, including all other absorbers and Tropic convection, is $\cong$ 0.16. The overall absorption metric is (1-Tcomb) and is $\cong$ 0.84. Approaching any asymptote, such as (1-Tcomb) approaching 1.0 — full absorption — is quite difficult. 0.84/(1 - 0.75)=3.4. Increases in CO_2 beyond 380 ppm only very slowly reduce the *eff*TCO_2 of 0.75.

Each K(0) surface temperature *requires* a specific Blanket flux at Z=0. A solution for the equilibrium K(0) is determined only when modeled subtractions, upwards slice by slice, from the required Blkt(0) are found to match the additive results for the downward flux. The downward flux is calculated from the cutoff altitude above which temperature and low density essentially eliminate further adiabatic effects. The selected adiabatic cutoff is Z=13.13 km (Z'=0.82), but results are fairly insensitive to any selection above Z=11.0 km. Composites of all power fluxes show the smooth dependencies on normalized altitude (**Figure 9**).

A calibration plot of equilibrium surface temperature K(0) vs. (1-TOth) for an *idealized* solo absorber Oth provides the fundamental comparison for any combination of actual absorbers. Oth is defined to have the same transmission TOth for all WN, where WN is photon wavenumber (cm^{-1}) and is directly related to frequency. (There would be no "notching" if the only absorber were an ideal Oth.) Oth is an abbreviation for *Other*, and, in combination with known absorbers, is often used to represent an unknown background (bkgnd) of absorption. Multiple examples are presented, and the calibration plot always holds up when a properly defined Tcomb replaces a solo (and ideal) TOth. Tcomb=TCO_2•TH_2O•...•Tbkgnd. (See **Figure 22**,) Relations between cumulative adiabatic extractions and cumulative post-adiabatic BBR absorptions are also presented. For example, the summations of all adiabatic extraction (up to the limit of Z'cutoff) are consistently equal to about 140 W/m² (watts per square meter). P(Z'cutoff) approximately equals 99 W/m², and there is only very limited absorption beyond Z' cutoff of the remaining BBR.

A better understanding of Tcomb, CO$_2$(WN), and the sensitivity of TOth

Transmission varies with temperature, but the tabular TMol(WN) for a specific molecule Mol with a specific ppm refers to a slab of appropriate thickness at a uniform 300K Standard Temperature and Pressure (STP) with no other absorbing molecules. The actual transmission and absorption for the entire atmosphere is determined from the slice-by-slice calculations. The tabular T(WN) is appropriately adjusted for the temperature and thickness of each slice. Published tables of T(WN) for various concentrations of Mol=CO$_2$ and Mol=H$_2$O exist.

The important conclusion is that each K(0) requires a specific value for the idealized (1-TOth). More specifically, each value of (1-Tcomb) for combined absorbers has the same K(0) as the matching (1-TOth) for a solo ideal absorber. "Ideal" TOth means that TOth(WN) is the same for all WN.

Since TCO$_2$(WN) have already been published for all WN for a set of CO$_2$ ppm, the "effective" overall TCO$_2$ encompassing all WN is the TOth that gives the same total absorption.

Published tables of T(WN) exist for CO$_2$={252.8, 632, 1,264, 2,528, 6,320, 12,640} ppm. Spreadsheet calculations for 252.8 ppm and 632 ppm have been interpolated to obtain a K(0) value of 288K for 380 ppm of CO$_2$. This baseline point requires the inclusion of an additional background uniform absorber with a TOth equal to 0.2140. *Eff*TCO$_2$ for 380 ppm≅0.758 and an averaged Tcomb=(0.758)•0.214)=0.1622. Therefore, (1-Tcomb)=0.8378. Consequently, any Tcomb of 0.1622, or a solo Oth with TOth=0.1622, gives a K(0) of 288K.

With the assumption that the same "background" TOth=0.2140 be maintained for higher and lower CO$_2$ values, it is determined that ΔK(0) for a doubling of CO$_2$ within the range {252.8, 632} is ~0.42C. {*If there were no background*, i.e., TOth=1.0, the maximum possible ΔK(0) for a doubling of CO$_2$ within the range {252.8, 632} would be 0.95K, and all K(0) would be well below 288.0K.} Any introduction of new molecules "sharing" the absorption reduces the contributing absorption by CO_2.

If TOth=Tbkgnd=0.214 is maintained for the next doubling, $\Delta K(0)$ for quadrupling is ~0.90C. If justified arguments can be given for accompanying decreases in Tbkgnd (more absorption) as CO_2 increases, the $\Delta K(0)$ values will increase (**Figures 3, 24**). Such arguments are discussed, but any arguments that link Tbkgnd to TCO_2 are tenuous. What would be helpful is the existence of two (or more) baseline points that must be simultaneously fitted.

When both TCO_2(WN) and TH_2O(WN) are included in simulations, the *eff*Tbkgnd will differ for the two studies since Tbkgnd for the CO_2-only case includes H_2O as background. But K(0) will be unchanged for the same Tcomb.

The truly important parameter is $\Delta K(0)$ per $\Delta(1$-Tcomb$)$. The rate increases with K(0), but an approximation (0.68 ± 0.01) for temperatures from 288K to 290K, is:

$$\Delta K(0) \cong 0.68C \cdot 100 \cdot (\Delta(1-Tcomb))$$

$(\Delta(1$-Tcomb$)) \cong (0.0062)$ for $\Delta K(0)=0.42C$. The surface Power P(0) increases from 390.08 W/m² to 392.36 W/m². The blanket, Blkt(0), increases by 2.28 W/m².

Further doublings (quadrupling, octupling, etc.) of CO_2 can be calculated. Quadrupling of CO_2 from 380 to 1,520 ppm will give $\Delta K(0)=0.90C$. With Tbkgnd retained at 0.214, 4,800 ppm are necessary to have $\Delta K(0)=2.0C$. Over 11,000 ppm is required for $\Delta K(0)=3.0C$.

These results may seem to be a surprising representation of reality, but the calculations hold up. What they also suggest is that Tbkgnd for $CO_2=252.8$ ppm should be higher than 0.214 (less absorption) and that Tbkgnd for $CO_2=632$ ppm should be lower than 0.214 (more absorption).

Since there is no defined second baseline point, official data on temperature rise over time was (loosely) correlated with official CO_2 rise over time. That apparent value for recent data is 0.28K/100ppm. (The author is not confident in the accuracy of these $\Delta K(0)$ and ΔCO_2 correlations, but they do allow a testing of how one can proceed if there were confidence in the data.)

With efforts unexplained here, the slopes were doubled from

the calculated 0.14K/100ppm to 0.28K/100ppm, and the ΔK(0) between the {252.8, 632} bookends were shown to be doubled from 0.528K to 1.056K. (See **Figure 20**.) 0.28K/100ppm is close to the observed data, and, on that basis, it might be claimed that 0.80K per doubling of CO_2 is a more accurate doubling factor. This is still well below the official predictions of 2.0 to 3.0K per doubling of CO_2, which, certainly, do not *backtrack* very well to CO_2 levels lower than 380 ppm. That alone makes them suspicious.

Orig:
{252.8, 287.821K, 0.214} →
　　　{380, 288.056K, 0.214} →
　　　　　(632, 288.349K, 0.214}
End-to-end ΔK(0)=0.528K → 0.40K/doubling;
End-to-end K(0) slope=0.14K/100 ppm

Mod:
{252.8, **287.856K, 0.2187**} →
　　　{380, 288.056K, 0.214} →
　　　　　(632, **288.542K, 0.2081**}
End-to-end ΔK(0)=1.056K → 0.80K/doubling;
End-to-end K(0) slope=0.28K/100 ppm
End-to-end ΔTbkgnd=-0.0106 →
　　±2.5% of median Tbkgnd has doubled the K(0) slopes

　　Although the result is significant, it doesn't solve the problem; however, it shows that the fitting problem can be solved. Small changes in Tbkgnd (5%) are much more effective than large changes in CO_2 (225%). CO_2 has been the chosen suspect. If an amended doubling effect of CO_2 were 0.8C per doubling, then ΔK(0)=3.0C would require over 5,000 ppm of CO_2.
　　Tbkgnd is almost totally due to water absorption and to Tropics-induced convection from water condensation. Exploratory simulation runs that try to produce the official prediction of 2.5K to 3.0K per doubling of CO_2 by aggressively reducing Tbkgnd because of the presence of water have been quite unreasonable. Although any rise in temperature, such as by CO_2 absorption, might be expected to

produce higher water vapor content in the atmosphere, any continual temperature rise associated with absorption by water will require more water to be *retained* over time. Rainfall "annihilates" the water vapor, and does not even guarantee the return, in the form of warm rain, of the heat *extracted* from the earth to create the vapor. (Just as CO_2 is increasing over time because its creation rate exceeds its annihilation rate, so too would the average water content have to be increasing.)

TPW and convection contributions to Blanket

The average integrated amount of water vapor in the atmosphere is defined by the term Total Precipitable Water (TPW). Literature results presented in Section Two indicate that water levels in the atmosphere, over many decades, have *not* been increasing. There is even TPW data as a function of latitude (**Figure 36**). The answer to the importance of water contribution should be resolvable from the analysis of reliable annual data. Total precipitation levels also do not appear to be increasing, although the variations with latitude (and, therefore, with all locations) indicate that "climate change" is a natural perturbation of nature. Even Relative Humidity (RH) has not been increasing; it has been *decreasing* (**Figure 33**). {This is not consistent with the data on TPW, but it is consistent with global observations of increased desertification. This study cannot answer this, but it suggests a totally different attack plan from saying that CO_2 is causing more water absorption in the atmosphere. Agricultural activities may be giving the sun more direct access to surface water. The cooling while the water is being depleted is replaced by direct surface heating after the surface water is depleted.}

The T(WN) tables for H_2O vapor actually are based upon TPW, rather than on ppm, because ppm content is not constant with altitude. (Average ppm's can be calculated.) Special changes were made to the workhorse simulation model to incorporate variations in TPW and convection-contributed blanket. In addition to having a spreadsheet column for $TH_2O(WN)$, integrations for absorption were limited to altitudes below $Z'=0.4$. Above this altitude, it is highly likely that vapor will be condensed and the realm of water

Global Warming Temperatures and Projections

condensation and heat injection into the atmosphere itself takes over. A very significant amount of the heat extracted from the earth has simply been transferred to higher altitudes and the total heat of condensation can be estimated. This is particularly true in the Tropics. This heat can be transferred as radiation, or as warm air convection. Even with higher than average TPW values of 5 cm and 10 cm (where cm represents the total amount of water if it all were condensed to a layer on earth), the *eff*TH$_2$OAbs values were approximately equal to that for CO_2. The calculated Tbkgnd becomes even more intolerable, and so an allowance is made for a variable entry of convection-contributed blanket to the computer model for the "average" state.

The available "free" blanket from the Tropics is considerable. The lowered 12 degrees at the surface correspond to about 70 W/m^2 of excess heat within that crown in the tropics. All of this is theoretically available to the remaining 60% of the earth's surface area. A positive Blkt(0) increment of 45 W/m^2 elsewhere corresponds approximately to a +8.0C addition to K(0). The Temperate — and to a small extent, the Polar areas — *will also be contributing* to the convection bus. None of this is incorporated in the Section One model except in so far as Tbkgnd can include contributions from the convection bus. For a mathematical resolution, the convection bus contribution must be eliminated from Tbkgnd. Multiple simulation runs were made with K(0)=288.349K, CO_2=632 ppm, *eff*TCO$_2$=0.7336, TPW=5 cm and TPW=10 cm, and with Conv (Tropics+local) contributions of 0 to 80 from the convection bus. Output values were *eff*TH$_2$OAbs+Conv, *eff*TH$_2$OConv, Tcomb, Tbkgnd. Solution sets that gave a final Tbkgnd greater than 0.85 were examined for their reasonableness. Whether for TPW=5 or 10, totally reasonable results were obtained for Conv=70 or 80, and (1-TH$_2$OAbs+Conv)/(1-TCO$_2$) varied from 2.8 to 2.9. (See **Tables 6, 7**). So if the Tropics contribution was, say, 40 W/m^2 and the local contribution were say, 30 to 40 W/m^2, there is nothing discomforting in the results. The Tropics contribution to K(0) in the Polar regions is estimated (by others) to be at least 10C, and so it is clear that the Tropics bus does not die out quickly.

Conclusions

The conclusions from this analysis are that H_2O is, indeed, the far greater contributor to K(0), that direct connectivity of CO_2 to greater H_2O absorption has not been demonstrated, that atmospheric convection from the Tropics significantly increases temperatures in the Temperate zone and decreases temperatures in the Tropics zone, and that CO_2 increases are in no way conducive to any likelihood of a runaway. Evaporation/evapotranspiration and subsequent condensation may contribute as much 50% of Blkt(0). (See the intitial table in the Quick references section that follows the appendicies.) Even very small reductions in reflectivity (albedo) can have a greater effect than CO_2 increases. And ocean currents — not discussed here — are also employed by nature to convect heat.

It is postulated (without proof) that the normally higher temperature rises over land may be further increasing as near-surface water is gradually reduced. This reduces net surface evaporation, reduces surface cooling and lowers atmospheric relative humidity.

The possibility of a "runaway" K(0) depends on the value of an appropriate "forcing function" for now and for all projections. In order for there to be a runaway, the combined net power of all forcing functions that can be expected to produce a K(0) rise of 1.0C must be less than the natural $\Delta P(0)$ for a K(0) rise of 1.0C. It takes an absorption of 10.8 W/m² to obtain a P(0) rise of 5.4 W/m² associated with a 1.0C increase in K(0). Any storm (weather change) is an attempt by nature not to accelerate disorder but to quickly restore equilibrium. The most peaceful weather days always occur after a storm.

Recommendations for improved modeling

- Establishment of at least one other baseline point.
- A total review of the present methods for establishing the definition of official daily temperature, calibration equivalence for all types of monitors, and new data for multi-year results that employ the same monitors for every year.

Global Warming Temperatures and Projections

- An improved understanding of K(0) and the sun in producing water vapor.
- A focused consideration on whether temperatures over land might be increasing (and RH decreasing) because of a gradual reduction of near-surface water and increased direct baking from the sun.
- Detailed modeling of the upper atmosphere heat bus.
- Transfer of this spreadsheet modeling to a mainframe computer.

Preliminary 'peeks'

The charts on the next three pages, generated from spreadsheet data, illustrate correlations between earth temperatures and atmospheric composition. These graphs are the result of calculations detailed in later chapters.

Figure 1 is the calibration plot for determining surface temperature as a function of whatever the net absorption is for any selection of absorbing molecules. ("Net absorption," as stated here and subsequently, is meant to include the convection increments to the blanket as specified in Section Two.)

Figure 2 is a localized magnification of the upper right of Figure 1. It displays the temperatures and levels of CO_2 in parts per million (ppm) that are of current interest.

Figure 3 illustrates a "doubling" calibration plot, which gives the surface temperature as the atmospheric CO_2 levels are doubled and when it is only CO_2 that increases. This figure is fully compatible with Figures 1 and 2.

22 Preliminary 'peeks'

Figure 1: K(0) vs. (1-Transmission)

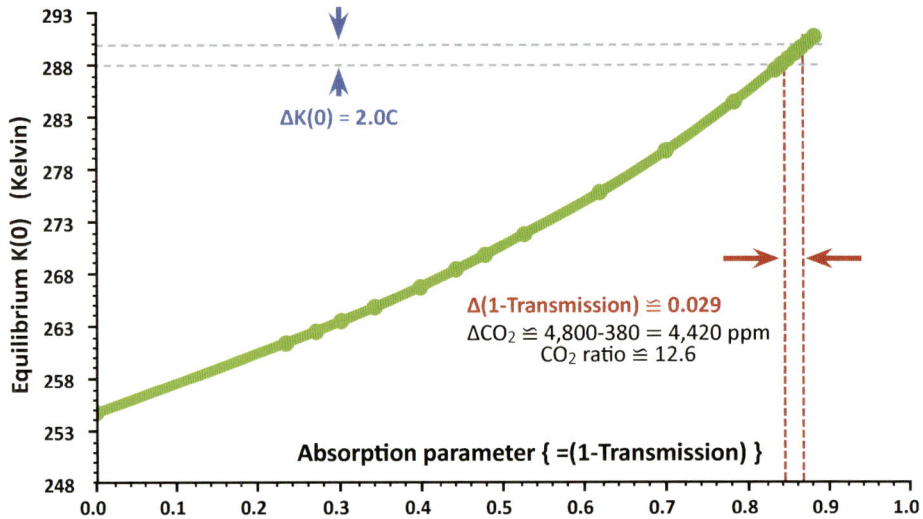

The average surface temperature of earth is related to the number of energy-absorbing molecules in the atmosphere. The current range is the top band: between 288° and 290°K (15°C to 17°C, or 59° to 63°F).

Figure 2: K(0) vs. (1-Transmission)

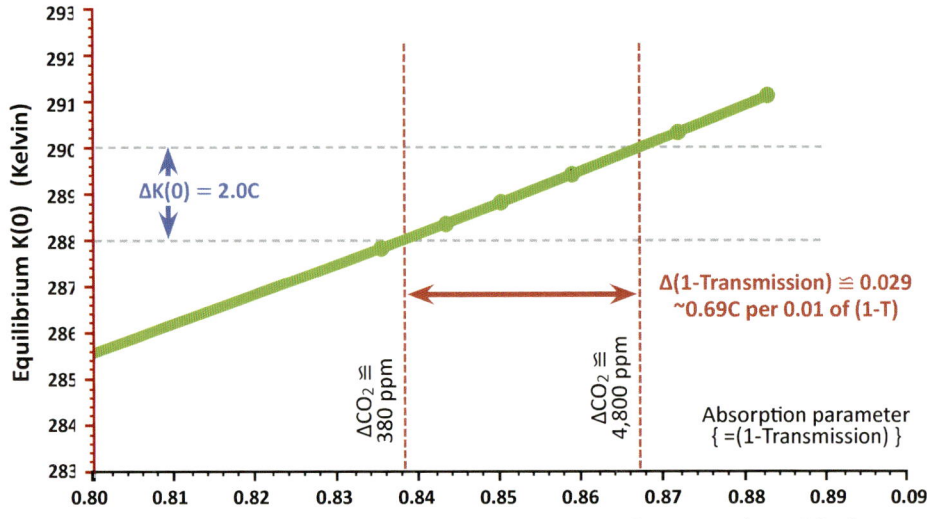

An amplification of Figure 1, showing how temperature changes when CO_2 in the atmosphere increases by a factor of 12.6.

Figure 3: $\Delta K(0)$ vs. $LOG2(CO_2/380)$ with $TOth = 0.214$

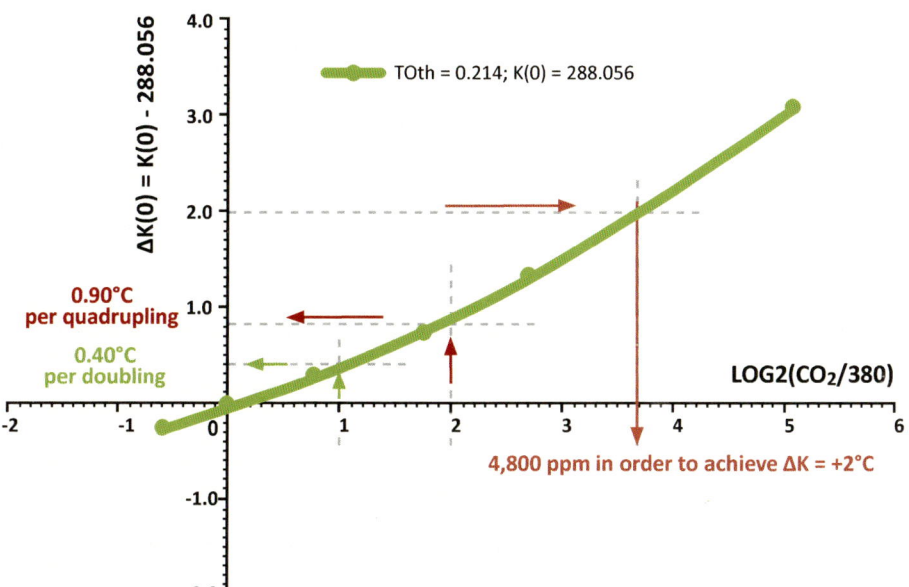

This third figure processes the data of the previous two in a different manner. Instead of concentrating on surface temperature as a function of a universal (1-Tcomb), it employs a fixed starting point and concentrates on surface temperature (change in surface temperature) as a function of increasing CO_2 relative to the fixed starting point. It is more meaningful to use a logarithmic scale for the CO_2 ratio and a LOG2 (rather than a LOG10) scale allows an easy inference of Delta temperature changes for each doubling of CO_2. Figures 1, 2 and 3 are in perfect accord.

Section One:

Modeling the atmosphere

Introduction

The (averaged) equilibrium of the earth's surface temperature depends on the dissipation/escape of each day's absorption of the sun's daily input. Increasing storage increases temperatures but, importantly, reduces nighttime drops. Each day's input collection and storage is dissipated since an overheated object emits a quantum mechanically defined Planck infrared black body radiation (BBR) towards the lower temperature direction.

The atmosphere and earth surface exhibit heat storage "charging" and discharging times, confirmed by the simple truths that June 21 is not the warmest day and December 21 is not the coldest day. Also sidereal noon is NOT the warmest time of day; and the hour before visual dawn is usually the coldest time of day. With less storage and/or days much longer than 24 hours, late night temperatures would be brutal.

The atmosphere demonstrates particular phenomena. Both the molecular density and the temperature decrease with altitude. The heat rise is not a molecular flow, but heat rise has an effect parallel to that of a pressurized gas flowing through a nozzle into a region of lower pressure. Expansion through a nozzle produces cooling because of the thermodynamic "work" associated with the expansion.

The direct parallel is a puff or air parcel of heat at the surface; as the heat unit rises it will expand and will cool. No new heat is added, but its initial heat suffices for the work of expansion. An extremely extended surface of a continuous heat source is simply restoring the height-dependent pressure and temperature that has already been established.

The atmospheric heat rise has an easy accommodation between pressure and volume such that the *cooling conforms its temperature to its surroundings*. Since there is (nearly) no work, i.e., no heat loss to the surroundings, the thermodynamic cooling is an "adiabatic cooling" effect.*

The sun's average globally absorbed input is 239 W/m^2. *If there were no absorption* anywhere in the atmosphere, the adiabatically-reduced remainder at 13 kilometers of the surface BBR of 239 W/m^2 would be about 100 W/m^2. The surface temperature K(0) would be 255K (-18C) — actually higher, since the formation of ice must be considered. The adiabatic factor over this range is 139/239, equal to 0.58.

When the infrared BBR is absorbed its "dissipation/escape" is complicated. No energy is lost because of the absorption. Half of the post-absorption re-emission can be considered to be directed back to the surface very much as if it were an addition to the sun's 239 W/m^2. This downward emission constitutes a "heat blanket." The temperature of the atmosphere and its K(0) automatically (thermodynamically) adjust to new higher temperatures. The downward emission has effectively experienced a warming adiabatic effect as the atmosphere seeks its new equilibrium.

The downward emission is reclaiming the adiabatic loss it had suffered as a directed BBR. This version of an explanation is based on an assumption the surface was not already in equilibrium. With the surface already in equilibrium then the downward half of any thermalized re-emission would be experiencing this anyway, i.e.,

---

* See, e.g., http://farside.ph.utexas.edu/teaching/sm1/lectures/node56.html. These are class notes prepared by Prof. Richard Fitzpatrick of University of Texas-Austin.

with no measurable flow.

This document determines the necessary (averaged) conditions to achieve equilibrium. All of the post-absorption emission has become thermalized and half of the thermalized radiation at any point in the atmosphere is always extending downward. Since at thermal equilibrium there is always a half up and half down situation there cannot be said to be a defined "directed" radiation downwards. However, it can be mathematically convenient to consider "up" and "down" as long as both are treated properly. And even in thermalized equilibrium there is always a net cooling adiabatic effect in the upward direction.

Nearly half of the entire thermalized radiation created from the absorption is the basis for *the* "blanket." The entire absorption contribution to thermalized radiation is continually adjusted with altitude for the adiabatic effect and can be simultaneously integrated from Z=0 to Z to obtain an "AdjCumSup(Z)" (which maxes near Z=4 km, and then falls off as the adiabatic depreciation dominates the collection). So too a RevAdjCumSdown from Zmax to Z=0 produces Blanket(0).

The new P(0) is approximately:

239+(RevAdjCumSdown(0)+AdjCumSup(0))
=239+Blanket(0))+0

The "adjustment" is, typically, a gain factor of ~1.3.

Half of any increase in the Absorption is an increase in Blanket(0). An increase in the blanket will have produced an increase in the surface BBR(0)=P(0) and each 5.4 Watts/m^2 increase in blanket corresponds to about a 1.0K (or 1.0C) increase in surface temperature K(0). Studying these increases is the crux of this document.

A "forcing function" could be described either as "Each 5.4 W/m^2 in blanket causes a 1.0C rise in surface temperature," or as "Each ~10.8 W/m^2 in atmospheric absorption causes a 1.0C rise in surface temperature."

Absorptions close to the surface exhibit low warming adiabatic gain but the lesser absorptions at upper altitude exhibit higher gain. Either results in a full "reclamation," with the total typically claiming a gain of ~1.3. P(Z) and K(Z) are automatically adjusted (by nature) to obtain the correct values for average $K(0)$ and $P(Z)$ distribution.

There is no thermal runaway. This is a false scare message. There is also nothing special about the absorption by CO_2 molecules. Physicists have modeled all the absorption modes of this linear O-C-O molecule. H_2O and other molecules have their own preferred wavelengths of absorption. CO_2 does not have its own special runaway-inducing properties. Existing H_2O molecules are, in fact, producing an approximately equal absorption as the CO_2 molecules.

Definitions

σ   Stefan-Boltzmann constant, $5.67 \cdot 10^{-8}\,W/m^2/K^4$, Watts per square meter per fourth power of degrees Kelvin. It relates the overall power at any temperature (or altitude) to the Kelvin temperature at that altitude. $P = \sigma \cdot K^4 = 5.67 \cdot (K/100)^4$.

k_B Boltzmann constant = $1.381 \cdot 10^{-23}$ Joule-sec/K; relates the kinetic energy of any gas molecule to temperature.

h   Planck constant. The Planck Constant is a physical constant and is the "quantum of action." Among other uses, it links the energy of a photon with its frequency. $h = 6.626 \cdot 10^{-34}\,m^2\,kg\,s^{-1} = 6.626 \cdot 10^{-34}$ Joule seconds. (Joule is energy, Joules/second is power, Joule-seconds is "impulse.")

C   Temperature in degrees Centigrade.

K   Temperature in degrees Kelvin, where K=273+C.

STP   Standard Temperature and Pressure; air at 300°K at sea level.

ppm   Parts per million. In this document, ppm are transformed into total CO_2 molecular count per square meter and Stull et al transmission $T(CO_2, WN, 300K)$ is determined for a

Definitions

Figure 4: Trans. Coeff. for 632 ppm CO_2, Trans. = 0.50 for Background "Other," and Planck Distribution for 288.0K

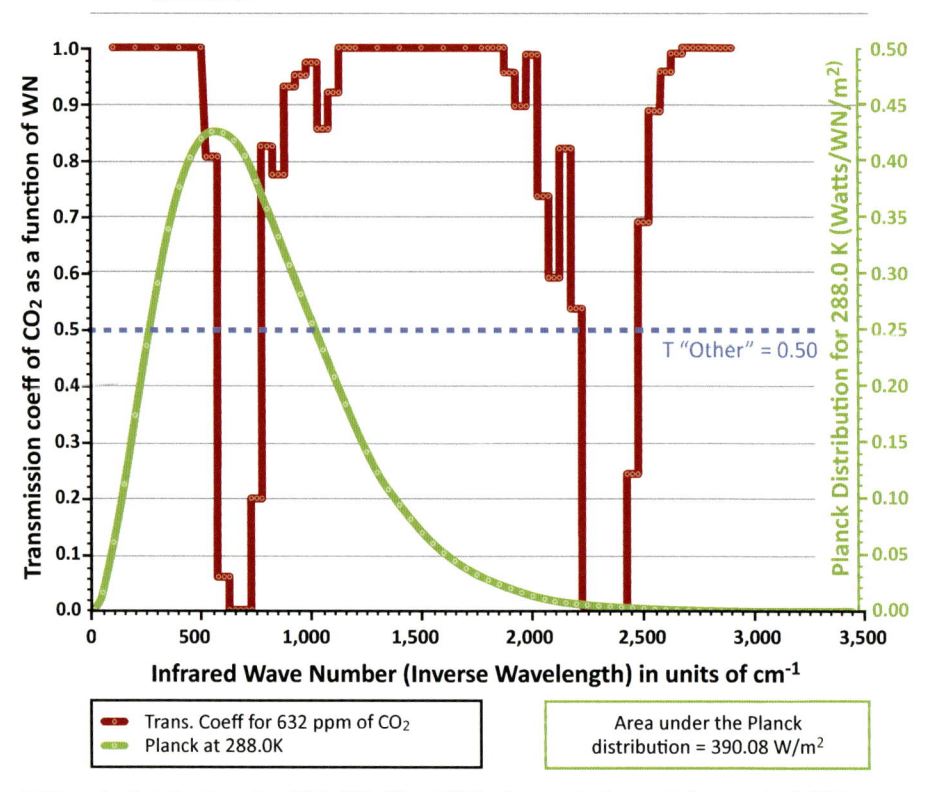

[Graph: Transmission coeff of CO_2 as a function of WN (left y-axis, 0.0 to 1.0) and Planck Distribution for 288.0 K (Watts/WN/m²) (right y-axis, 0.00 to 0.50) vs. Infrared Wave Number (Inverse Wavelength) in units of cm⁻¹ (x-axis, 0 to 3,500). Dashed line at T "Other" = 0.50.]

Legend:
- Trans. Coeff for 632 ppm of CO_2
- Planck at 288.0K
- Area under the Planck distribution = 390.08 W/m²

A Planck distribution for 288.0K. The TCO₂ factor is for a CO_2 level of 632 ppm. There are two obvious spans of WN where CO_2 is particularly sensitive to absorption, but there is little power flux in one of those spans. At higher altitudes (lower K) the Planck power will be lower.

---

	uniform slab at a temperature of 300K.*
BBR	Black body radiation. The electromagnetic radiation of a non-reflective object in thermodynamic equilibrium with its environment.
Z	Atmospheric altitude in units of kilometers (km). Half of all molecules in the atmosphere are below 5.3 km.

Global Warming Temperatures and Projections 33

Z'	Normalized altitude from 0.0 to 1.0. Its scale measures the integrated fractional count of molecules at that altitude.
W/m²	Watts per square meter, or power flux, usually presented as a function of wavenumber, WN.
J	Joule. A Joule is a unit of energy. Its units are W-s. That is, one Watt of power dispenses one Joule of energy per second. Or one Joule every one 3,600th of an hour
WN	Wavenumber. WN is the inverse of wavelength, with the wavelength given in units of centimeters.** (See X-axis of **Figure 4**.)
P(Z')	The overall Power per square meter (W/m²) of radiation at the altitude Z' (or Z). It is equal to the sum of the (directed) BBR power and the (isotropic) thermalized power. (P(Z')=BBR(Z')+2• CumAdjSup(Z).)
T	The transmittance factor, PowerOut/PowerIn. The maximum value of T is 1.0, which occurs when there is no absorption.*** (See left hand Y-axis of **Figure 4**.)
T(WN)	The transmission coefficient as a function of WN, or the PowerOut(WN) for a unitary PowerIn(WN)=1.0 for each WN. The transmission factors T(WN) are formally defined as transmission at 300K STP, with no adiabatic effects.
T(WNmid)	Transmission factor for the mid-value of each ΔWN. T(WNmid) is derived by integrating over ΔWN, and dividing by ΔWN.**** (See left hand Y-axis of **Figure 4**.)
Tcomb	Product combination of the T transmission factors of all molecules in a gas, and, typically, has product dependencies on WN, Tcomb(WN). The lowest TMolecule, the strongest absorber, dominates over the other absorbers.
ΔWN	Narrow bands of WN. In this document each ΔWN band is equal to 50 inverse centimeters.
Smid	In this analysis, the atmosphere is divided into slices with equal numbers of molecules and the interactions among adjoining slices are examined. Two Smid are defined for each slice ij. Smidij is one half the post-absorption emission from that slice. One Smid becomes an eventual component of CumAdjSup; the other Smid becomes a component of RevAdjCumSdown. These fluxes continue to have

Blkt(Z') Blanket radiation, the downward directed thermalized radiation, W/m², that eventually defines Blkt(0). P(0)=239 (the solar input)+Blkt(0) (the downward-directed post-Absorption emission).

Planck distribution

In short, a Planck distribution is the equilibrium power flux distribution from an (ideal) object or gas at temperature

---

* The percentage of CO_2 in the atmosphere is ppm CO_2/10,000. The actual count of all molecules after integrating upwards through the atmosphere is about 0.2 million trillion trillion ($2.0 \bullet 10^{29}$) per square meter. 400 ppm corresponds to about 80 trillion trillion per square meter. And over an area of 1.0 square Angstroms (about one tenth the area of a molecule), 400 ppm corresponds to 800,000 molecules per A^2. (It is very difficult for photons at particularly sensitive WN to escape without 99% of them being absorbed at very low altitudes.) 380 ppm of CO_2 compacted to a single slab at STP, with no other molecules present, would have a thickness of 300.6 centimeters. It is officially defined as a thickness of 300.6 atmosphere-centimeters (atm-cm). The Stull et al tables for 100, 200, 500, 1,000, 2,000, 5,000, and 10,000 atm-cm correspond, respectively, to 126.4, 252.8, 632, 1,264, 2,528, 6,320, and 12,640 ppm. {1.0 ppm=0.791 atm-cm; 1 atm-cm=1.264 ppm}.

** A higher WN corresponds to a higher frequency of radiation. For example, an infrared wavelength of 15 microns has a WN of 667 cm^{-1} (spoken as "667 inverse centimeters"), and a frequency of 20 trillion cycles per second. This is a particularly sensitive regime for absorption by CO_2.

*** The dimensionless T values from Stull et al are for slabs of CO_2 at atmospheric pressure and 300K. Furthermore, the table entries are for integrations over specified bands of $\Delta WN=50$ inverse centimeters. Linewidth broadening and overlaps are accounted for by Stull et al. In their application to pressures no greater than atmospheric pressure, pressure effects are negligible. In the modeling results presented here all transmission T(WN, CO_2 ppm) temperature effects are accounted for by modifying the T(WN) table to the actual Planck distributions (and thicknesses) for each atmospheric slice.

**** V. Robert Stull, Philip Wyatt, and Gilbert N. Plass, "The Infrared Transmittance of Carbon Dioxide," Applied Optics, Vol.3 issue 2, pp 243-254 (1964). The calculations include all quantum mechanical absorption lines. WNmid is the mid WN for a span of $\Delta WN = 50$.

K. It can be presented as the differential power per unit of WN (or wavelength or frequency). Integration over all WN gives the total power flux associated with that temperature. The total Power density increases as K^4: $P(0) = 5.67 \cdot ((K(0)/100)^4$. (See right hand Y-axis of **Figure 4**.)

Pli, Plj, Plmid
Planck distribution values as a function of WN at the Z'i, Z'j, and Z'mid values for each slice. The Planck values are for the Ki, Kj, and Kmid temperatures for these Z'. They are crucial inputs for calculating the ultimate WN dependencies of absorption and transmission and for identifying absorption notches.

A very brief summary of the spreadsheet model

In order to calculate how the sun's energy affects the earth's surface, it's necessary to understand how energy moves through the atmosphere. Energy is not lost, but some energy is doing adiabatic "work."

However, because the earth's atmosphere is not uniform it cannot be treated as a single entity. The atmosphere changes in temperature and density continuously with altitude.

To analyze the atmosphere as a whole, it is necessary to analyze what is happening everywhere at once, and then integrate the results to determine the overall equilibrium.

This document divides the entire atmosphere into thin slices and creates a summation of how each layer interacts with the ones above and below.

The spreadsheet model used for these calculations incorporates the following:

A very brief summary of the spreadsheet model

- The earth's altitude axis is normalized so that it goes from 0 to 1.0 rather than 0 to infinity kilometers.
- The atmosphere is divided into 50 slices, each with its own temperature and known count of CO_2 molecules.
- The earth's radiation spectrum is partitioned into 100 distinct bands of wavenumber.
- Proper models of absorption and net transmittance are applied within each slice.
- Published data of CO_2 transmittance as a function of inverse wavelength are employed.
- The natural thermodynamic "adiabatic cooling" as a function of altitude is included.
- All boundary conditions are carefully matched.

The approach can be compared to a standard Schwarzschild* transform except that it:
- Employs transmission, rather than absorption, parameters.
- Sums, rather than integrates over wavenumber.
- Integrates absorption and adiabatic effects.

An assumed surface temperature combined with the absorption in each slice dictates what the downwards "thermal blanket" at each altitude must be, but the modeling also calculates, top to bottom, what the thermal blanket actually is.

Iterations of surface temperature provide a verified solution.

The calculations demonstrate that for the present conditions of temperature and ppm of CO_2, there must be additional absorption by molecules other than CO_2. The absorption by that solo background absorption factor is three times greater than for solo CO_2. If one begins with the background absorption, the additional absorption by CO_2 contributes only 8% of the actual total absorption.

If all other absorbers remain unchanged, CO_2 would have to increase by more than a factor of ten in order to have a surface temperature increase of +2.0C.

---

* See www.barrettbellamyclimate.com. It discusses Schwarzschild's equation in an appropriate differential mode.

Normalization of altitude

Instead of slicing the atmosphere into layers that are the same depth in kilometers, these calculations divide the atmosphere into layers that have the same molecular density. In this way, there is a transformation of altitude Z in km to a normalized (dimensionless) Z'.

A complete description of the altitude transformation from 0 to infinity in kilometers to a normalized Z' is given in **Appendix B**.

While looking at the following graphs, here are some numbers to keep in mind:
- Half of all the molecules in the earth's entire atmosphere are below 5.308 km. For this reason, 5.308 km is defined as Z'=0.5. An exponential distribution* with Z is then fitted to define Z' for all Z.
- Z' is the integrated fractional molecular count at the corresponding altitude Z. Z' (in km) = -LN(1-Z')/0.13058538.
- ΔZ'=Z2'-Z1' is equal to the fractional molecule count between any two Z' altitudes.
- P(Z) and K(Z) are accordingly transformed into P(Z') and K(Z') relations.

---

* A molecular distribution more acutely inclusive of the adiabatic effect is considered later in the document.

Normalization of altitude

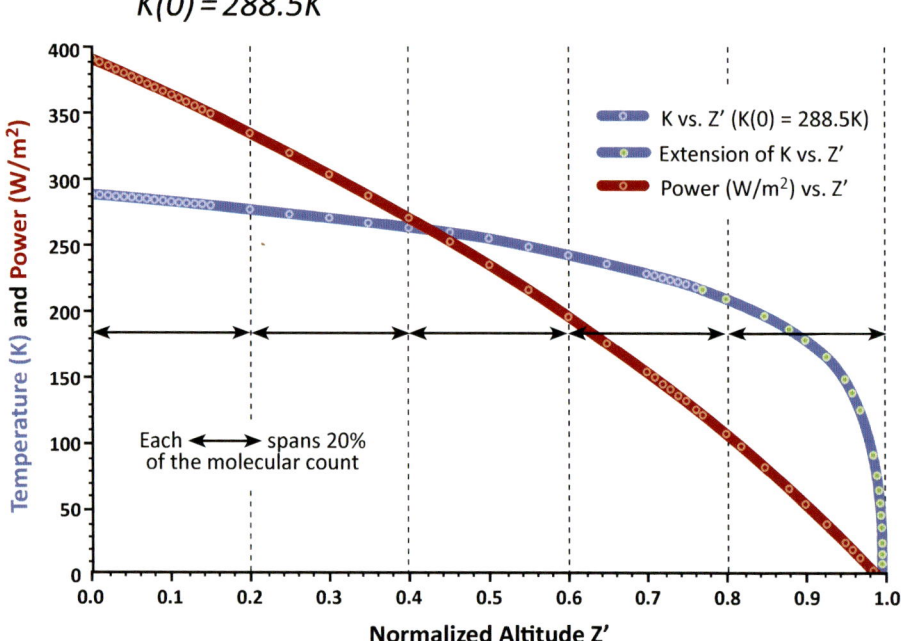

Figure 5: Temperature K vs. Normalized Altitude Z':
K(0) = 288.5K

[Figure 5: Plot showing Temperature (K) and Power (W/m²) vs. Normalized Altitude Z'. Legend: K vs. Z' (K(0) = 288.5K); Extension of K vs. Z'; Power (W/m²) vs. Z'. Annotation: "Each ↔ spans 20% of the molecular count"]

The blue plot shows how atmospheric temperature, measured in degrees Kelvin, and projected as if there were no ionosphere effects, falls drastically at high altitudes. The red overall power plot falls more smoothly, but it also ignores special atmospheric effects beyond Z'=0.82 (13 km). At all altitudes $P = 5.67 \cdot (K(Z')/100)^4$.

- **Figure 5** shows P(Z') and K(Z') for a surface temperature of 288.5K.
- These relations are formally modified whenever K(0) changes.
- The atmosphere is sliced into ΔZ' layers. The points on the P(Z') and K(Z') plots correspond to slice edges.
- Typical slice widths are 0.01 (1%) to 0.05 (5%). Continuity across boundaries is rigorously assured.
- A complete description of the altitude transformation from 0 to infinity in kilometers to a normalized Z' is given in **Appendix B**.

Basic BBR, power and Planck expressions

Before providing the details of the equations within each slice, a description of the basic equations for overall Power, BBR (black body radiation) power, and Blanket flux is worthwhile. Overall Power at any altitude is the sum of the (directed) BBR power and the (isotropic) thermalized flux produced when the BBR flux is absorbed. The Blanket flux is a component of the thermalized flux; it provides heat to the surface that is supplementary to that provided by the sun's input. The surface returns the blanket to the atmosphere as BBR.

Here are some straightforward observations. Please note that the transmission parameter is concerned only with the reduction of the BBR with altitude. Each slice will have its own transmission parameter, and their concatenated product provides the overall BBR transmission factor.

- Calculations in this document are averages over hours of the

Basic Expressions

day, months (seasons) of the year, latitude and longitude, as in Trenberth.[*]

- At Z=0, Z' also equals 0. The averaged (upward directed) infrared:

$$BBR(0) = BlackBodyRadiation(0) = P(0)$$

$$P(0) = \int_{WN=0}^{WNmax} Pl(WN,0) \bullet dWN = \sum_{WNmin}^{WNmax} Pl(WNmid,0) \bullet 50$$

The Planck Power distribution $Pl(WN, K(Z)) = Pl(WN, Z)$ is the power density (in Watts per square meter) per wavenumber WN (inverse of wavelength in units of cm^{-1}). Integrating over all WN at any specified K(0) gives the total BBR(0) in W/m^2. This integrated result always gives the Stefan-Boltzmann result:

$$BBR(K(0)) = 5.67 \bullet (K(0)/100)^4 = \sigma \bullet K^4 = 5.67E\text{-}8 \bullet K^4$$

where $\sigma = 5.67 \bullet 10^{-8}$ (the Stefan-Boltzmann constant).

$$Pl(WN, Z') = \frac{\pi \bullet 2E8 \bullet h \bullet c^2 \bullet WN^3}{e^{100 \bullet h \bullet c \bullet WN/(k_B \bullet K(Z'))} - 1}$$

where h=Planck constant, k_B=Boltzmann constant, c=speed of light and where the Planck distribution Pl(WN, Z') has units of $W\ m^{-2}\ WN^{-1}$.

For all other Z':

$$BBR(Z', WN) = T(Z', WN) \bullet (Pl(0, WN))$$

where (1-T(Z',WN)) is the fractional reduction of the original BBR, and includes all absorptions and all adiabatic loss.

A crucial issue is that absorption must be calculated at each WN. The final remaining (remanent) BBR power flux is an integral (or summation) over all WN.

---

[*] Trenberth et al, http://www.cgd.ucar.edu/staff/trenbert/trenberth.papers/BAMSmarTrenberth.pdf.

Calculations within each slice

With the atmosphere divided into slices containing equal numbers of molecules, the next step in the analysis is to mathematically describe what is occurring in each slice.

What follows is a basic explanation. An accompanying schematic (**Figure 6**) is a simplification of a more detailed figure in **Appendix C**.

1. The first consideration in the calculations for each atmospheric slice is that it is only the absorption of the directed infrared BBR from the surface $Z'=0$ that is of consequence in producing any effects on any ultimately downward "blanket" flux. Thermalized flux is non-directional (isotropic); both its pre-absorption WN distribution and its post-absorption WN distribution are identical, That is, the definition of thermalization says that its emissions mimic its absorptions.

2. A second consideration is the adiabatic effect. In the upwards (cooling) direction there is an adiabatic loss for both BBR and for

(both halves of) the thermalized flux expanding ("adiabatically") at all altitudes greater than Z'=0. The adiabatic effect vanishingly diminishes at altitudes beyond which molecular spacing delays molecular interaction. A "cutoff" of adiabatic attrition is usually ascribed to a particular altitude.

3. A third consideration is that absorption is very much dependent upon WN, and so a BBR distribution can be drastically altered from its Planck distribution at Z'=0. The thermalized flux will always rapidly restructure so that, mutually, the total distribution matches the power uniformity required by the local temperature.

4. A fourth consideration is that all BBR absorptions will, overall, result in thermalized flux. Half of the BBR absorption initiates the formation of the blanket.

In the formation of the blanket all of the pre-absorption adiabatic loss is restored and the ultimate Blkt(0) is greater than the half of the (post-adiabatic) absorption that created it. That is, Blkt(0) is greater than Abs/2 and the following ratio is observed in all situations.

$$Blkt(0)/CumAbs \cong 0.64.$$

5. A fifth consideration, an extension of 4, is that the theoretically downwards flux effectively experiences a negative — i.e., a heating — adiabatic effect. All of the (theoretically) lost flux due to the cooling adiabatic effect on the upwards BBR is fully restored. That is, half of the adiabatic dissipation is restored. The fact that only half of the absorbed BBR contributes to surface temperature is, itself, a significant deterrent to any "runaway" model. On the other hand, the very fact there is an absorption inhibits the ability of the surface to allow a daily "escape" of the sun's input.

6. A sixth consideration is that the relative adiabatic effect (relative to P(0)) decreases as P(0) increases but the total adiabatic effect is unchanged. The total adiabatic loss at any altitude Z is 239-P(Z). At the cutoff altitude Z = 13.13 km, the adiabatic loss is always about 140 W/m² (see **Figures 29** and **30**).

In general, there is a strong tendency towards stability, and no tendency towards runaway effects. If CO_2 increases by a factor of 4, total absorption increases by only 9%, blanket power increases by

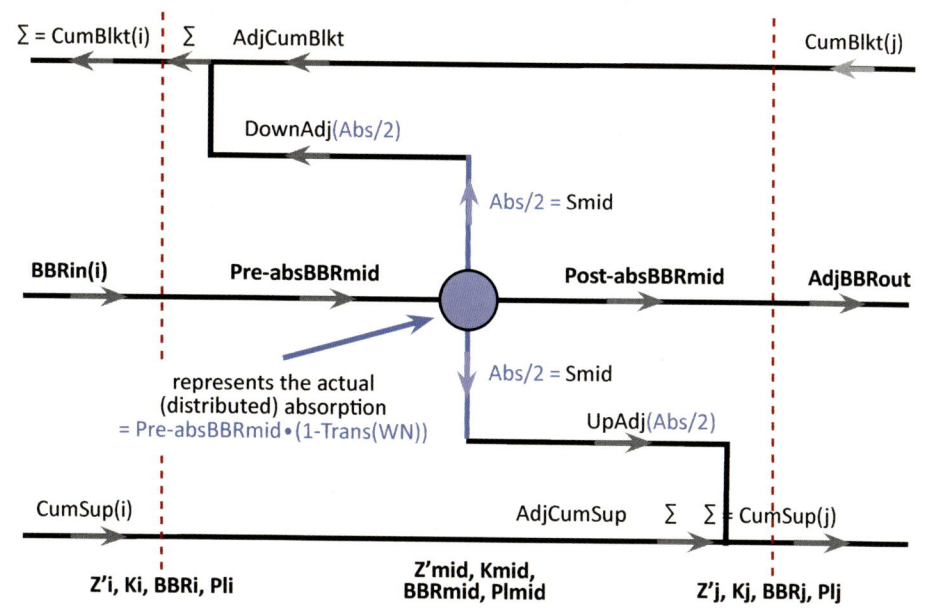

Global Warming Temperatures and Projections 45

Figure 6: *Interactions within each atmospheric slice*

Σ = CumBlkt(i)   Σ AdjCumBlkt   CumBlkt(j)

DownAdj(Abs/2)

Abs/2 = Smid

BBRin(i)   Pre-absBBRmid   Post-absBBRmid   AdjBBRout

represents the actual
(distributed) absorption
= Pre-absBBRmid • (1-Trans(WN))

Abs/2 = Smid

UpAdj(Abs/2)

CumSup(i)   AdjCumSup   Σ   Σ = CumSup(j)

Z'i, Ki, BBRi, Pli   Z'mid, Kmid,   Z'j, Kj, BBRj, Plj
                          BBRmid, Plmid

A schematic of the interactions to be tracked within each slice: BBR absorption, cumulative upwards adiabatic reduction, and cumulative downwards growth of the "Blanket."

only 6%, and the ratio of P(0) to total absorption decreases by 5.5%. That's quite the opposite of a runaway.

There is an orderly process for evaluating each slice. All calculations are carried out for one WN band at a time — and, eventually, summed. (See accompanying schematic for slice ij. **Figure 6**) If there were no absorption and only an adiabatic effect, the slice-averaged BBR would equal its adiabatic-reduced value midway into the slice. It is this pre-absBBRmid that is multiplied by the WN-dependent absorption transmission parameter for this slice width at the Planck temperature Kmid. This produces a post-absBBRmid.

BBRin experiences adiabatic loss and becomes pre-absorptionBBRmid. After absorption the BBR flux is reduced to postabsorptionBBRmid. Before it exits at Zj' it has experienced more adiabatic loss. This second adiabatic loss is actually a component of pre-absorption adiabatic loss for slice jk, the next slice.

$$(1-T(WN)) \cdot (preabsBBRmid(WN))$$
$$= [preabsBBRmid(WN) - postabsBBRmidWN]$$
$$= 2 \cdot Smid$$
$$= 2 \cdot Absorption/2$$

Smid in the upper half of the schematic represents the post-abs thermalized emission that is directed downwards. For spreadsheet bookkeeping reasons the WN dependence is retained until the time of merger with the downwards blanket flow across the top.

DownAdjSmid(WN)

$$= \left[Pli(WN) \Big/ Plmid(WN) \right] \cdot (Abs/2)$$

$$= (1/2) \cdot (1-T(WN)) \cdot (BBRin(WN))$$

The blanket component in each slice is a direct giveback of one half of the absorption.

In the lower half of the schematic Smid represents the post-absorption thermalized emission that is directed upwards. Once again for spreadsheet bookkeeping reasons the WN dependence is retained until the time of merger with the upwards blanket flow across the bottom.

The upward AdjSmid(WN)

$$= \left[Plj(WN) \Big/ Plmid(WN) \right] \cdot (Abs/2)$$

$$= \left[Plj(WN) \Big/ Pli(WN) \right] \cdot (1/2) \cdot (1-T(WN)) \cdot (BBRin(WN))$$

The total absorption, 2 • CumSmid, is determined for the equilibrium K(0) associated with each choice of CO_2 and TOth. 2 • CumSmid at any Z' minus the value of 2 • AdjCumSup along the parallel highway gives the adiabatic losses for the thermalized power.

Global Warming Temperatures and Projections

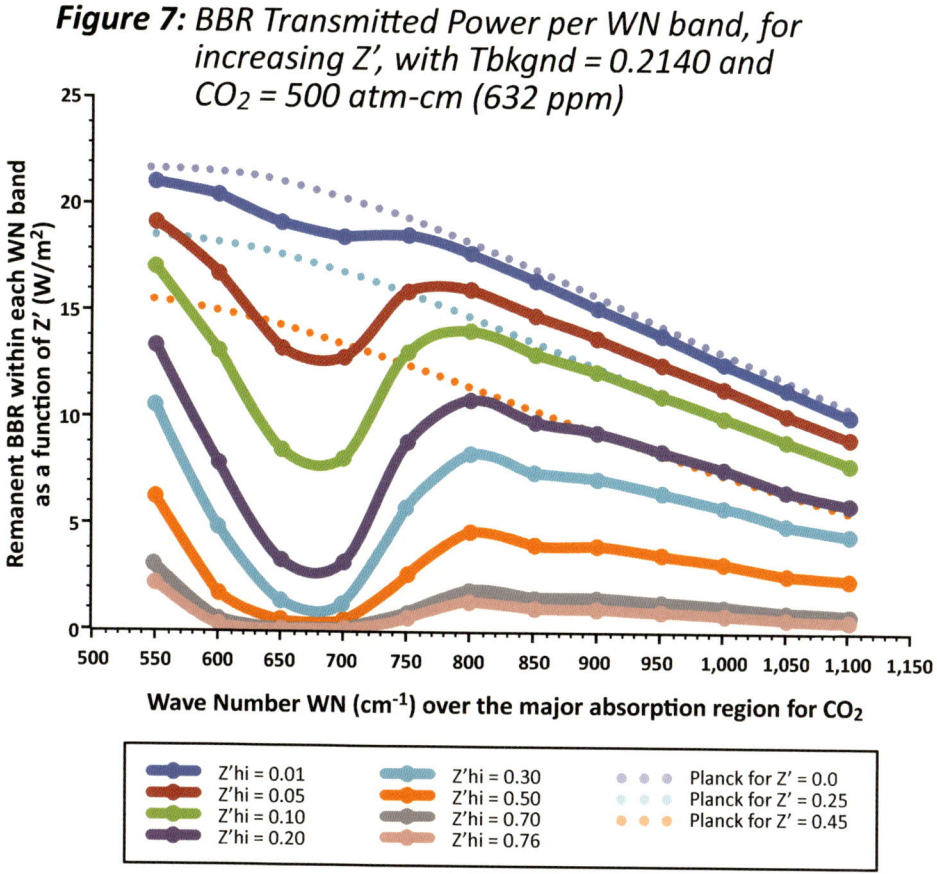

Figure 7: BBR Transmitted Power per WN band, for increasing Z', with Tbkgnd = 0.2140 and CO_2 = 500 atm-cm (632 ppm)

Y-axis: Remanent BBR within each WN band as a function of Z' (W/m²)

X-axis: Wave Number WN (cm⁻¹) over the major absorption region for CO_2

Legend:
- Z'hi = 0.01
- Z'hi = 0.05
- Z'hi = 0.10
- Z'hi = 0.20
- Z'hi = 0.30
- Z'hi = 0.50
- Z'hi = 0.70
- Z'hi = 0.76
- Planck for Z' = 0.0
- Planck for Z' = 0.25
- Planck for Z' = 0.45

In this plot of energy absorption in an atmosphere at 632 ppm of CO_2, the notch illustrates that all wavelengths are not absorbed equally. If the atmosphere had no absorbing molecules, the BBR distributions at different altitudes would match those of the three dotted line examples. Nearly all of the absorptions beyond the widest notch are being caused by the uniform background absorption.

This schematic emphasizes symmetries between the upper Blanket track and a lower "AdjCumSup" track. But both Smid legs connect to a 2 • AdjCumSmid track. The entire thermalized distribution 2 • AdjCumSmid is susceptible to adiabatic loss. The Blanket track is

Calculations within each slice

Table 1: Values at equilibrium where $CO_2 = 632$ ppm

Property	Value
P(0)	391.97 W/m²
Blkt(0)	152.98
CumBBR absorption* (Z'=0.82)	236.83
CumBBR transmission loss, w/o adiab	236.83
CumBBR adiabatic loss** (0.82)	134.92
Cum(Δ(-BBR)) loss (0.82), incl adiab	371.76
Altitude to achieve a full "notch"	~5 km
"Additional" net abs. because of CO_2	~7%
Remaining BBR transmission (0.82)	20.0 W/m²
2 x AdjCumSup (0.82) (thermal)	75.86
* Ratio; CumAbs(0.82)/P(0)	0.604
** Ratio: CumBBRadiab(0.82)/P(0)	0.344

Summary values for an equilibrium condition with CO_2=632 ppm, background absorber with a uniform transmittance of 0.2140, and K(0)=288.349K.

---

operating on only a downward component, AdjCumBlkt. The blanket track represents the "fill" for the absorptions so the proper increase in P(0) can be realized. There is no actual flow at equilibrium.

In summary, all components add up to the correct bookkeeping sums for BRR transmission:
- An exact P(Z') at each slice edge:

    (AdjBBROut+2 • AdjCumSmid)

- BBR pre-absorption adiabatic losses.
- All thermal post-absorption adiabatic losses.

The spreadsheet correctly equalizes the slice boundaries. A separate test, which split a slice into five slices, gave the same output results to three significant figures.

Two related sets of bookkeeping are not shown. The required

Blanket(0) is known a priori for any K(0). For each slice the actual contribution to the blanket is calculated and each increment, properly adjusted for adiabatic effects, is subtracted and is set as the requirement for the next slice.

There is also an additive column of AdjCumSdown entries from Z'max to 0.0 that is tracked within the spreadsheet. Actual calculated contributions to the blanket are inserted into that highway and, at each slice boundary, the sum has a reverse adiabatic gain assigned. The results should be closely matching the upward subtractive column. At Z'=0, the result for the additive column must equal the required blanket. K(0) is adjusted until it does. The spreadsheet is sensitive to K(0) values ±0.001K.

Because all calculations are carried out over narrow (50 cm^{-1}) wavenumber bands, the development of "notches" in the BBR distribution is clearly demonstrated.

One such example is given in **Figure 7**, but more information is given in **Appendix D**. In this example, for 632 ppm of CO_2, a notch is fully established halfway up into the molecular count (Z'=0.5). The original BBR distribution is shown in the uppermost plot with dotted lines.

After each slice has been summed over WN, **Table 1** gives some of the important parameters associated with Z'=0 and Z'=0.82.

A second representation of the notch

In **Figure 7**, each data point in the uppermost dotted light purple line represents the Planck flux at Z'=0 within each range of WNmid ±25 cm^{-1}. The area under this entire plot, including not-shown extensions to the left and right, is P(0). P(0) is reduced to about 20 W/m^2 at Z'=0.82. So the ultimate transmission factor for the BBR is 20/P(0)≅0.05 for the range of P(0)'s being considered. This is relatively independent of CO_2 and TOther, except to the extent that P(0) changes when the absorption changes.

At Z'=0.82, P(0)-20 has been "lost" to absorption and to the adiabatic effect, with the sum total of the pre-absorption adiabatic to absorption ratio being about 0.57. This ratio decreases somewhat as the total absorption increases.

The dotted teal line in **Figure 7** corresponds to a Planck distribution for temperature K at an altitude of Z'=0.25. The dotted orange line corresponds to a Planck distribution for K at an altitude of Z'=0.45. It's easy to see how the original BBR is being devoured as Z' increases, even halfway through the molecular count.

Figure 8 shows the overall (dimensionless) transmission coefficient for BBR with altitude as Z' increases. The losses from WN=875

Figure 8: BBR Transmission factor vs. WN, including adiabatic loss, from Z' = 0 to slice output {BBR T averaged over WN ±25 cm^{-1}; CO_2 = 500 atm-cm (632 ppm); TOth = 0.2140}

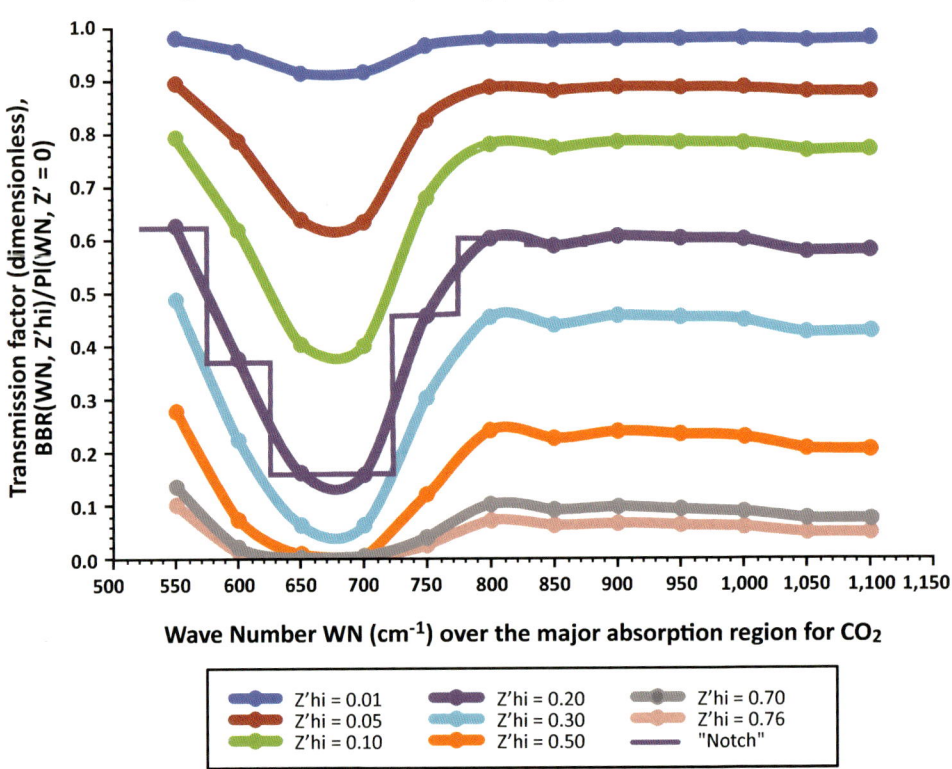

Transmission factor (dimensionless), BBR(WN, Z'hi)/PI(WN, Z' = 0)

Wave Number WN (cm^{-1}) over the major absorption region for CO_2

Z'hi = 0.01
Z'hi = 0.05
Z'hi = 0.10
Z'hi = 0.20
Z'hi = 0.30
Z'hi = 0.50
Z'hi = 0.70
Z'hi = 0.76
"Notch"

With the same input data as Figure 7, this figure presents the (dimensionless) transmission factor for the remaining BBR at a selection of altitudes, relative to the BBR power at Z=0. "Notches" develop at CO_2-sensitive wave numbers, and remain bottomed out as CO_2 levels increase. The wave numbers primarily affected by the background are near saturation at the higher altitudes. At Z'=0.82 (13 km) only 5% to 6% of the initial BBR(0) remains.

to 975 are almost entirely due to TOth, a powerful absorber. CO_2, within its limited range, gets its job done quickly but the background Oth is steady and relentless.

The overall picture of atmospheric interactions

Previous figures and the appendices to follow provide concepts of: Planck BBR (black body radiation) from heated surfaces; atmospheric absorption; absorption "notches"; adiabatic effects; the essential requirements of a "blanket"; and a normalized 0 to 1.0 Z' axis for altitude that replaces the 0 to infinity kilometer Z axis.

Overall atmospheric power fluxes and temperature as a function of Z' have been demonstrated. Various sub-elements of the atmospheric fluxes have been described, with particular emphases on BBR devolvement, BBR transmission (and absorption) with multiple absorbers, and an integrated slice model for representing the atmosphere. It is a fundamental requirement that the sun's daily 239 W/m^2 input must escape or in some way be eliminated. Other than direct escape to outer space, the only other way to dispense with this energy is by means of adiabatic "work" within the atmosphere. Absorbed energy is never a "loss," but the half of the absorption that

54   *The overall picture of atmospheric interactions*

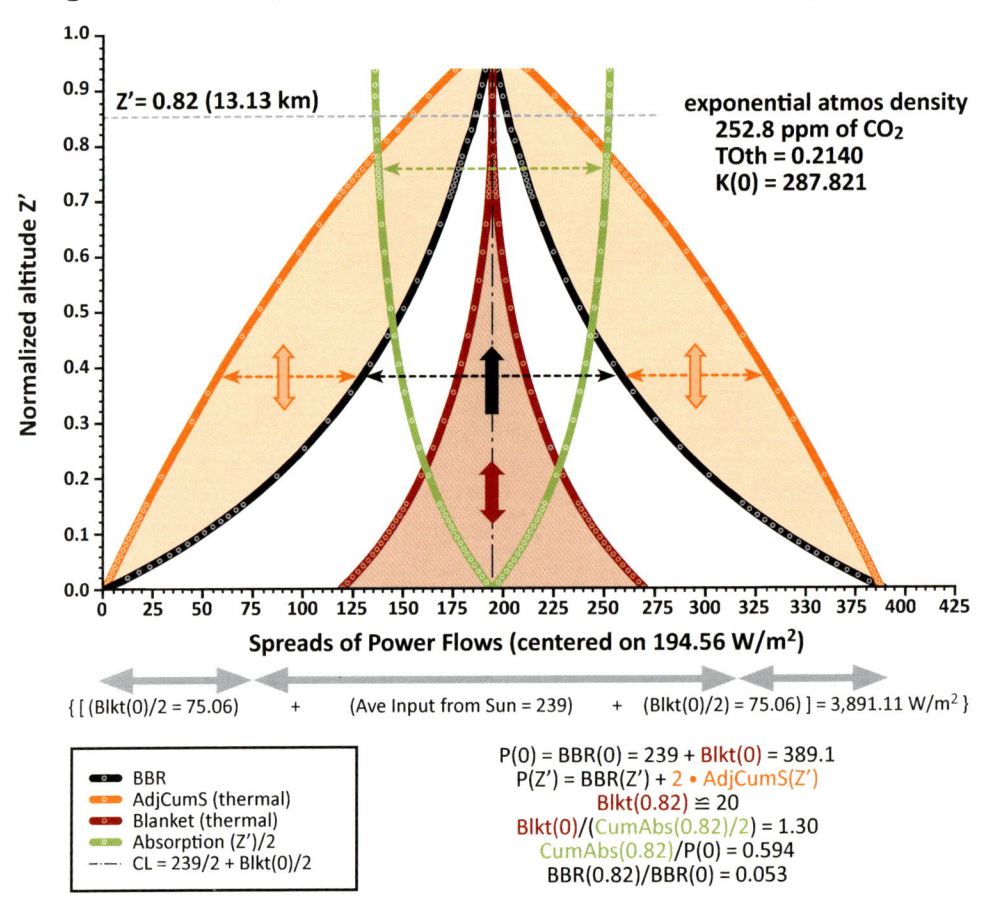

Figure 9: Atmospheric Equilibrium with $CO_2 = 252.8$ ppm

$Z' = 0.82$ (13.13 km)

exponential atmos density
252.8 ppm of CO_2
TOth = 0.2140
K(0) = 287.821

Normalized altitude Z'

Spreads of Power Flows (centered on 194.56 W/m²)

{ [(Blkt(0)/2 = 75.06) + (Ave Input from Sun = 239) + (Blkt(0)/2 = 75.06)] = 3,891.11 W/m² }

- BBR
- AdjCumS (thermal)
- Blanket (thermal)
- Absorption (Z')/2
- CL = 239/2 + Blkt(0)/2

P(0) = BBR(0) = 239 + Blkt(0) = 389.1
P(Z') = BBR(Z') + 2 • AdjCumS(Z')
Blkt(0.82) ≅ 20
Blkt(0)/(CumAbs(0.82)/2) = 1.30
CumAbs(0.82)/P(0) = 0.594
BBR(0.82)/BBR(0) = 0.053

As energy radiates upward and downward through molecules in all layers of the atmosphere, an equilibrium is reached. In this chart, the vertical axis is the normalized altitude Z'. Horizontal left to right separations at each altitude indicate the power flux strengths for: **upward-directed BBR**, **thermal downward expanding Blanket**, **isotropic thermalized radiation**, and one-half the **cumulative sum of all absorptions** from the upward-directed BBR that become thermalized.

eventually comprises the blanket warms the surface and allows the additional adiabatic work from the immediate BBR generation to satisfy nature's requirements. The atmosphere has become an

Global Warming Temperatures and Projections

energy storage "battery," fluctuating from day to night.

More absorption necessarily dictates a higher surface (averaged) equilibrium temperature K(0).

The described slice model (here and in the appendices) describes the exact mathematical process for determining these (averaged) equilibriums.

Before proceeding to the final "calibration" plots, it is best to summarize the full results for two specific sets of calculations. One is for 252.8 ppm of CO_2 and the other is for 632 ppm. This is a ppm ratio of 2.5 that straddles the range of ppm of current concern. It is calculations such as these that both create the calibration plots and confirm the calibration plots, plots which will be shown to extend past 12,000 ppm.

The first figure, (**Figure 9**) is for CO_2=252.8 ppm, a TOther of 0.2140, and K(0) of 287.821. The second (**Figure 10**) is for CO_2=632 ppm, TOther = 0.2140, and K(0) = 288.349. Each is an equilibrium solution. The difference in surface temperature for this ratio of 2.5 in CO_2 is 0.528C. This corresponds to a linear ΔK(0) of 0.40C per "doubling" of CO_2.

There are four color enclosed regions in **Figures 9** and **10**, including one which has two isolated parts (shown this way only in order to emphasize understanding).

The X-axis is a power axis (which is generally shown as a Y-axis, but NOT here). The Y-axis is the normalized Z' axis. The left to right spans across a colored region give the power flux at that altitude. So what is seen, in general, is the depletion of power with altitude, and the apportionment of power with altitude.

The lateral spread of the black "tent" is the BBR with altitude. With altitude it reduces because of absorption and because of pre-absorption adiabatic effects.

Orange is for the thermalized radiation, both up and down, produced by absorption. (However, the orange regions do not include the thermalization associated with the blanket.) Both the left and right edges of the black figure have orange lines beneath. The orange under black line on the left side is to be correlated with the orange line to the farther left. The orange under black line on the right side is to be correlated with the orange line to the farther right.

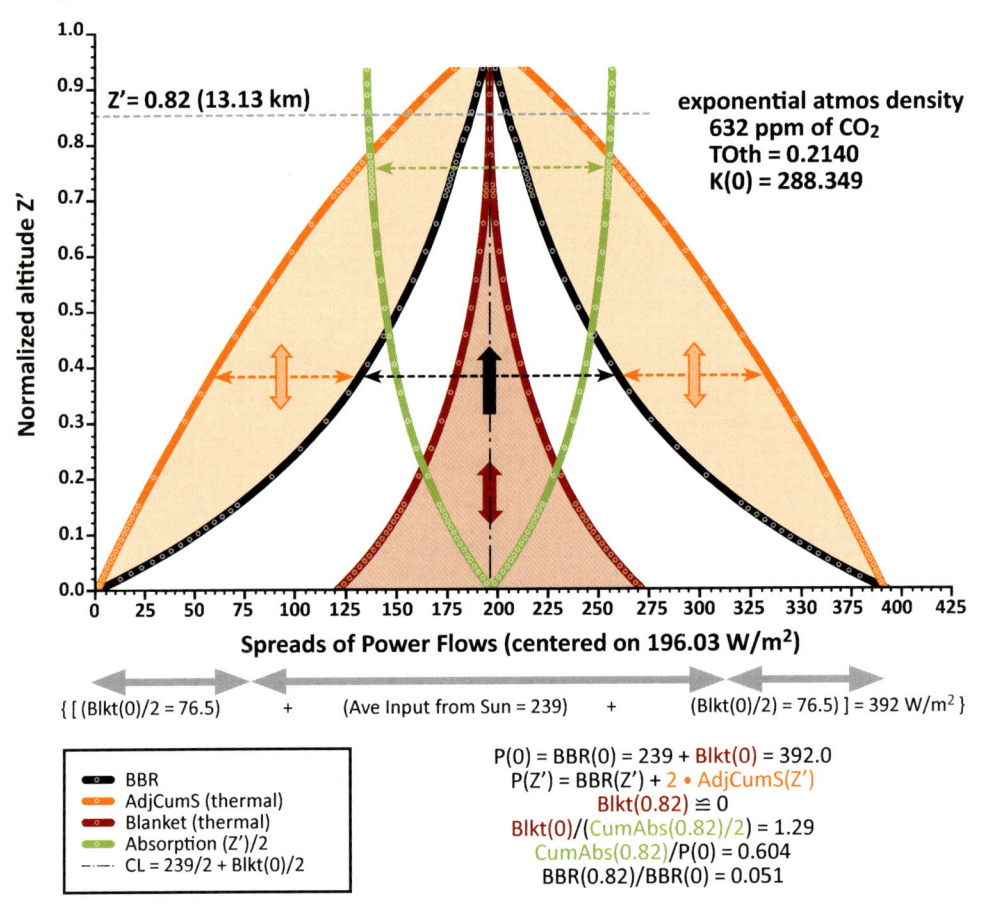

Figure 10: Atmospheric Equilibrium with $CO_2 = 632$ ppm

$Z' = 0.82$ (13.13 km)

exponential atmos density
632 ppm of CO_2
TOth = 0.2140
K(0) = 288.349

Normalized altitude Z'

Spreads of Power Flows (centered on 196.03 W/m²)

{ [(Blkt(0)/2 = 76.5)   +   (Ave Input from Sun = 239)   +   (Blkt(0)/2) = 76.5)] = 392 W/m² }

- BBR
- AdjCumS (thermal)
- Blanket (thermal)
- Absorption (Z')/2
- CL = 239/2 + Blkt(0)/2

P(0) = BBR(0) = 239 + Blkt(0) = 392.0
P(Z') = BBR(Z') + 2 • AdjCumS(Z')
Blkt(0.82) ≅ 0
Blkt(0)/(CumAbs(0.82)/2) = 1.29
CumAbs(0.82)/P(0) = 0.604
BBR(0.82)/BBR(0) = 0.051

When the amount of carbon dioxide increases, a new equilibrium is reached. With 2.5 times the amount of CO_2 as in Figure 9, absorption increases by 1.9%, K(0) increases by 0.528 degrees, and P(0) increases by 2.86 W/m² (0.74%).

At any altitude Z' the sum of the spans across the two orange areas plus the span across the black area is equal to P(Z'). (This, of course, is the same as the span across the two outer orange lines.) Since P(Z') = BBR(Z')+2 • AdjCumSup, **Figure 9** gives a perfect confirmation of this.

The BBR is the only constituent at Z'=0 and covers the entire allocated span of the X-axis. That span equals the 239 W/m² sun

Global Warming Temperatures and Projections

input plus the Blanket(0) input. It is directed upwards. At Z'=0.82, the remanent BBR is close to 20 W/m².

The red "tent" is the Blanket and develops downwards. It is very close to zero at Z'=0.82, and Z'=0.82 is chosen as the cutoff for the reverse adiabatic effect. The adiabatic effects above 0.82, whether positive or reverse, have no consequence to these calculations.

The two orange eyes (or "bananas") are not physically disassociated with altitude but are shown in this way because they add (laterally, in this presentation) to the BBR power at each altitude. They are the post-absorption creations of thermalized power. The combination represents the creation of thermalized radiation after each absorption minus the adiabatic effect it suffers with altitude. It has a maximum around Z'=0.45. It is directed both upwards and downwards. Its retained amount at Z'=0.82 is about 75-80 W/m². The sum of the two orange areas is more than double the total red area.

Figure 9 also supplies a visualization of another aspect of the "flows." The red tent, expanding downwards, is not really "flowing" as some independent flux, but is the thermalized blanket. Technically, it represents an expansion of the orange areas since it too is thermalized flux. The thermalized blanket flux is directly matched by a component of the upward BBR(0). This is directly displayed within the notations for the X axis. BBR(0)=239+Blanket(0). At higher altitudes of Z', the remnant BBR is compensated by lower values of the blanket and also by the downward portions of the post-absorption (and post adiabatic) thermalization in orange.

The span from green line to green line is not shown with any arrow direction. It is simply the cumulative sum of all absorptions from the upward-directed BBR that become thermalized. The cumulative sum is NOT present at each altitude; it is simply a recorded summation — no adiabatic effects, etc.

A challenge to the reader is to try to spot anything dramatically different between **Figures 9** and **10** for two CO_2 ppm that vary by a factor of 2.5. There are no dramatic shifts.

The only shifts are a shift of the centerline, a slide for the Z'=0 entries, and very small changes at Z'=0.82.

What is also interesting about these two sets of plots is that one (252.8 ppm) is close to the CO_2 level at the start of the "industrial

Figure 11: Atmospheric Equilibrium with $CO_2 = 632$ ppm incorporating adiabatic temperature changes

Z'= 0.82 (13.13 km)

Adiabatically adjusted atmos density
632 ppm of CO_2
TOth = 0.2139
K(0) = 288.349

Normalized altitude Z'

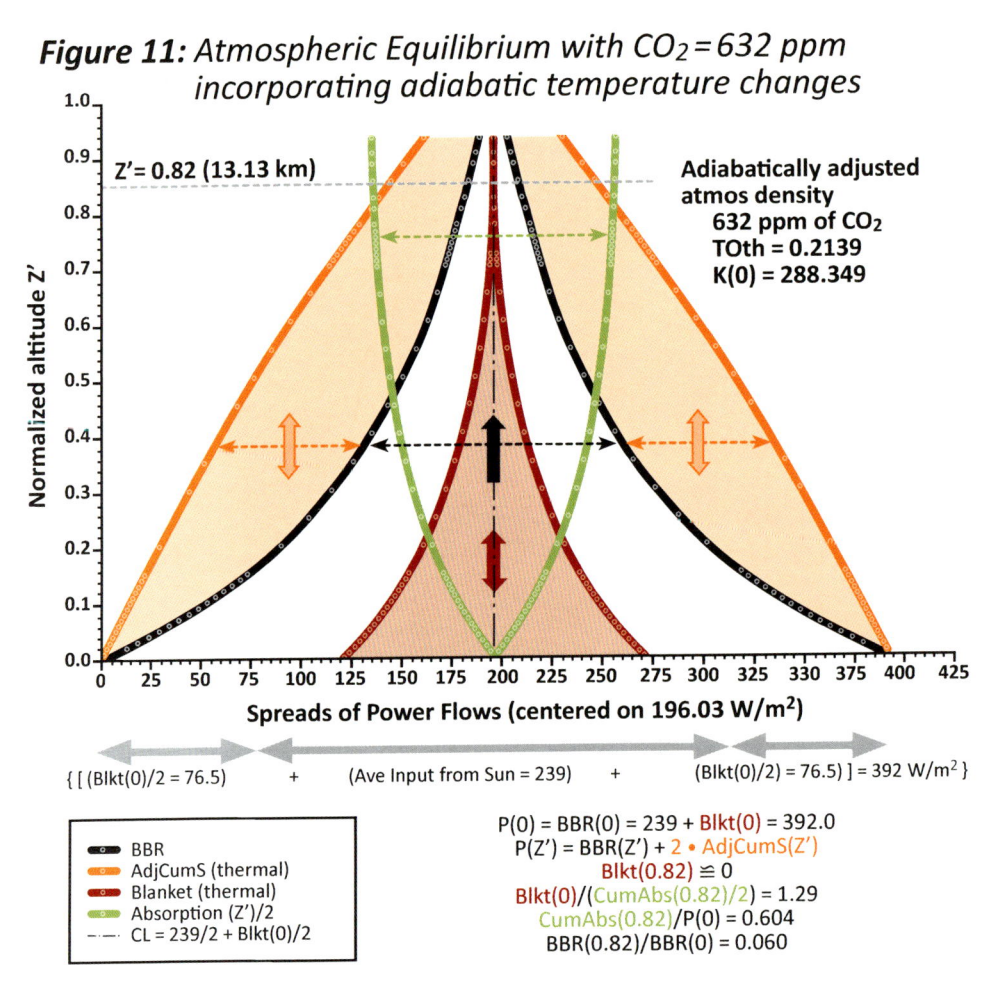

Spreads of Power Flows (centered on 196.03 W/m²)

$\{ [(Blkt(0)/2 = 76.5) + (\text{Ave Input from Sun} = 239) + (Blkt(0)/2) = 76.5)] = 392 \text{ W/m}^2 \}$

- BBR
- AdjCumS (thermal)
- Blanket (thermal)
- Absorption (Z')/2
- CL = 239/2 + Blkt(0)/2

P(0) = BBR(0) = 239 + Blkt(0) = 392.0
P(Z') = BBR(Z') + 2 • AdjCumS(Z')
Blkt(0.82) ≅ 0
Blkt(0)/(CumAbs(0.82)/2) = 1.29
CumAbs(0.82)/P(0) = 0.604
BBR(0.82)/BBR(0) = 0.060

A regeneration of Figure 10 to incorporate adiabatic temperature changes. The only significant differences are that Z'=0.82 occurs at Z=11.4 km and that the "escape" at Z'=0.82 is slightly greater, with BBR(0.82)≅25.2 and 2 • AdjCumS≅93.8. Their sum equals P(0.82) for K(0.82)=214.6K.

age" and the second could represent a situation 200 years from now (632 ppm). Yet there is nothing alarming, and the K(0) temperature increase over this 2.5 range in CO_2 is only 0.53C.

Molecular density and altitude

Previous calculations have employed an exponential drop-off of molecular density with altitude Z, with the fitting parameter that half of all molecules are below 5.308 km. And the K falloff has been linear with Z, which is the observational truth.

This section describes that drop-off mathematically in order to further later discussions.

The constraint of having an adiabatic falloff reduces the three parameters of the ideal gas law to two parameters.

The ideal gas law is pV=nRK, where p is the pressure of the gas, V is the volume of the gas, n is the amount of gas (in moles), R is the ideal, or universal, gas constant and K is the absolute temperature of the gas.

One such relationship with an adiabatic falloff is:

$K^{1.4}/p^{0.4}$ = a constant

The web site farside.ph.utexas.edu gives the following expression as its recommended representation of molecular density ρ with altitude.

For Z< Z1=20 km:

$$\frac{\rho}{\rho 0} = \left[1 - \frac{\gamma-1}{\gamma} \cdot \frac{Z}{z0}\right]^{\left[\frac{1}{\gamma-1}\right]}$$

For Z>Z1=20 km:

$$\frac{\rho}{\rho 0} = \left[0.714 \cdot \frac{Z}{8.4}\right]^{2.5}$$

The pressure p falls off more rapidly than K, and the molecular density falloff is intermediate between the two. The approximation above is not much different than the exponential falloff employed within the spreadsheet. This has been integrated to produce fractional molecular count Z', along with K(Z') and P(Z').

The most recently plotted results for 252.8 and 632 ppm of CO_2 were redone to obtain the same baseline requirement {380 ppm, 288.0K, TOth=?} with the new Z'. The new estimate for TOth was 0.2139, hardly a squiggle different from 0.2140.

Figure 11 employs the molecular density recommendations of farside.ph.utexas.edu. The absorptions are not occurring at the identical altitudes but the ultimate differences appear to be inconsequential. The spreadsheet, with some additional complication, can handle any modified Z'.

Soft entry to the calibration tables

Most of the following pages relate to calibration of K(0) to (1-Tcomb) and to calculations of ΔK(0) with doublings of CO_2.

The calibration plots of:

    K(0) vs. (1-Tcomb)

or of:

    ΔK(0)=K(0)-refK(0) vs. Δ(1-Tcomb)
    =[(1-Tcomb) for K(0) - ((1-Tcomb) for refK(0)]

are fixed for any set of absorbers, including or excluding CO_2.

A "doubling" plot of:

    ΔK(0) vs. $LOG2(CO_2/refCO_2)$

must define the tabular values and variations with CO_2 of all other absorbers in order to be an acceptable "calibration."

The first discussions define how a baseline point is established for the basic calibration. For example if the baseline demands:

{refK(0), refCO₂}

then it is quickly realized that the baseline demands an additional background absorption:

{refK(0), refCO₂, reqTOth}.

A "doubling" plot is quickly presented, but it (at least, initially) assumes that the required TOth for the baseline point applies to all other CO_2 levels as well.

The effects of a constant vs. changing TOth above and below 0.2140 are discussed. The doubling plots are modified but the overall (1-TOth) calibration plot is shown to be unaffected.

If there were two baseline demands, there would be less uncertainty about the progressive change of TOth. For example, if the two bases were:

{ref1K(0), ref1CO₂}
and
{ref2K(0), ref2CO₂}

two required TOth values could be determined:

{ref1K(0), ref1CO₂, req1TOth)
and
{ref2K(0), ref2CO₂, req2TOth}

It would be very desirable to agree upon two significantly separated baselines since interpolations (and some extrapolation) could be made between (and beyond) the TOth values.

Figure 12 in the following chapter gives the "doubling" plot for ΔK when TOth remains fixed at 0.2140 for all CO_2 levels.

Calibration

Author's Note: There is repetition of previous text in this section, but the intent is to make this chapter section self-sufficient.

Multiple plots are shown in the following sections. All are accompanied by a common (universal) calibration plot. Real (that is, calculated) data with equilibrium K(0) always lay smoothly on top of the calibration plot.

A previous section, *Calculations within each slice*, provides the guidelines for how these calibration plots are created. For any given CO_2, plus any other specified side conditions, K(0) is adjusted until a stable solution is obtained. Output data is saved to a designated storage area. New inputs are provided and those outputs are saved. Calibration plots are assembled.

To have a calibration table, there must be at least one baseline marker for a certain set of inputs that must produce a certain output. The baseline condition here is the joint achievement of {CO_2=380 ppm, K(0)=288.0K}

A K(0) of 288.0K cannot be achieved if CO_2 is the only absorber. CO_2 is a strong infrared absorber for certain wavenumbers, but a poor absorber at other wavenumbers. So the first conclusion is that CO_2 needs some level of additional support (molecules that will provide additional absorption).

The chosen backup is an ideal absorber (called "Other") that will absorb at any wavenumber, and will absorb equally well at every wavenumber. This is not an absolute requirement but is a logical choice as a "background" absorber. The only question then is how strong this absorber must be. So the new baseline condition is:

$\{CO_2 = 380\ ppm,\ K(0) = 288.0K,\ TOther = ?\}$

Calibration plots

The Stull et al tables that are closest to 380 ppm are for 252.8 ppm (200 atm-cm) and 632 ppm (500 atm-cm).

LOG2(252.8/380)=-0.588 and LOG2(632/380)=0.734; therefore, on a LOG2(CO_2/380) scale, LOG2(380/380)=0.0 is about 0.45 of the scale difference between 252.8 and 500.

A (spreadsheet) column for TOth(WN) = constant = TOth is placed alongside a column for TCO_2(CO_2=252.8 ppm, WN). A (somewhat) random value for TOth is chosen. The row entries are multiplied to obtain Tcombination(WN). TOth is adjusted to obtain an equilibrium value of K(0)≅287.8 as the spreadsheet output. For TCO_2(632, TOth), the hope might be to obtain a K(0) of 282.25. This is repeated until a comfortable solution is found for TOth such that a straight line between the two LOG2 points crosses the origin about 45% of the way between 252.8 and 500.

The solution is TOth=0.2140. The two temperature solutions are 287.765 and 288.293. The temperature difference is 0.528K (=0.528C); the CO_2 ratio is 2.5.

This transforms into a "doubling" statement:

The surface temperature increases by 0.40C for a doubling of CO_2.

It also requires (actually, assumes) that TOth=0.2140.

The doubling plot is given in **Figure 12**. The plot is extended to four additional CO_2 values: 1,264, 2,528, 6,320, and 12,640 ppm.

Global Warming Temperatures and Projections

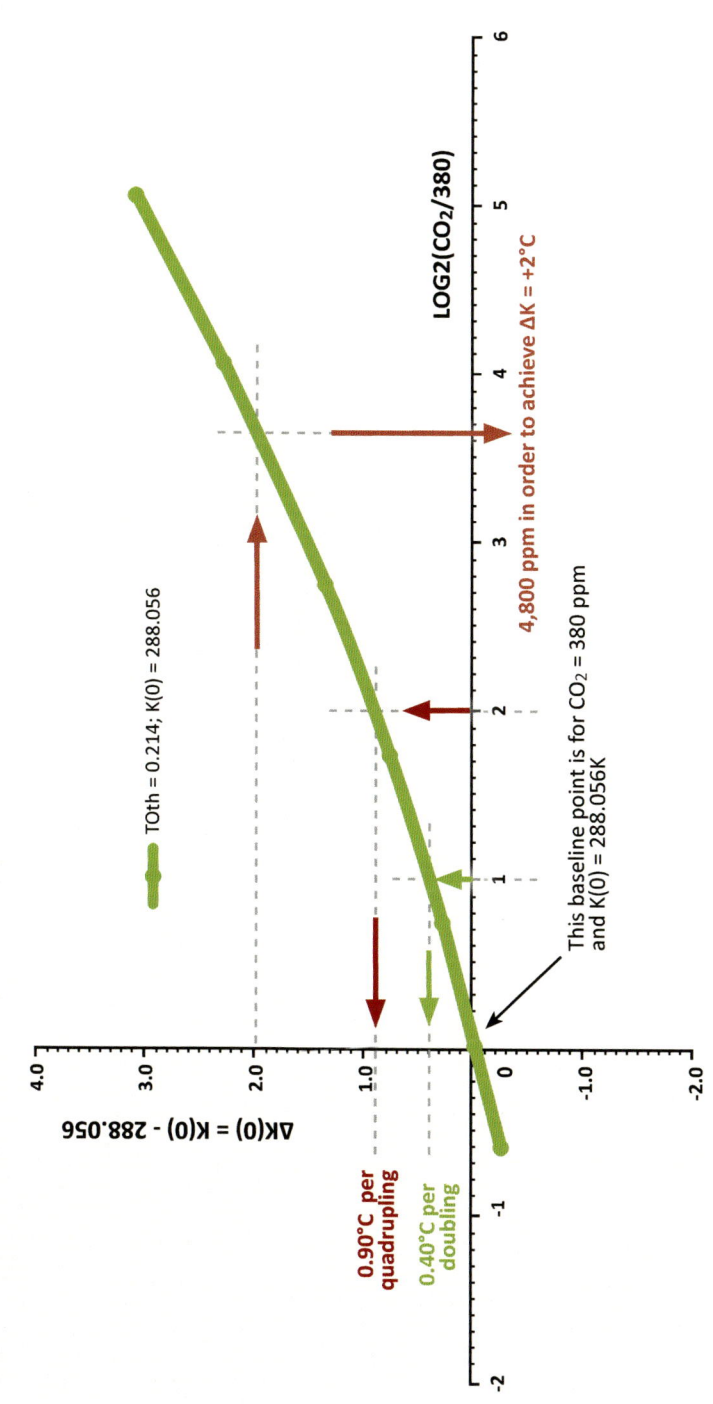

Figure 12: $\Delta K(0)$ vs. $LOG2(CO_2/380)$ with $TOth = 0.214$

A calibration plot on a logarithmic scale illustrates that each doubling of carbon dioxide results in an increased earth surface temperature of less than one half of one degree Centigrade. The plot was derived from measurements published by V. Robert Stull, Philip Wyatt, and Gilbert N. Plass in their 1964 paper in the scientific journal Applied Optics, "The Infrared Transmittance of Carbon Dioxide."

That last value is 12.6 ppk (parts per thousand). And the temperature increase is only 3.1K!

Terms used in K(0) calibration plots

T is the general term for the dimensionless Pout/Pin transmission coefficient. Each absorber will have its own T(WN).

TOther, however, is an idealized absorber that absorbs equally at all WN. (See **Figure 4** for TOther=0.5.)

(1.0-T) is the dimensionless absorption coefficient for each absorber.

For each CO_2 ppm there is a **ΔWN • TCO₂(WNmid)** value for each band of WN.

***eff*TCO₂** is the effective T over all WN for a specified ppm of CO_2, one that matches the overall effect of a TOth. That is, K(0) for TCO₂ of a specified ppm has the same overall transmission as for the general TOther producing the same K(0).

There can be a **TMol** for any other molecule (such as H_2O). "Other" can also refer to an unknown background of absorbers.

TOther in this case relates to an unknown background that is assumed to have the same transmission factor for all WN.

When two or more absorbers are present, their combined transmission factor is:

Tcomb(WN)=TA(WN) • TB(WN) • ...TOther

A particular Tcomb used for these simulations is:

Tcomb=TCO₂ • TOther.

Figure 13 is the calibration plot of K(0) vs. (1-TOth). The only absorber is the idealized "Other." The analysis determines the equilibrium K(0). The calibration plot equally well applies to (1-Tcomb).

Global Warming Temperatures and Projections

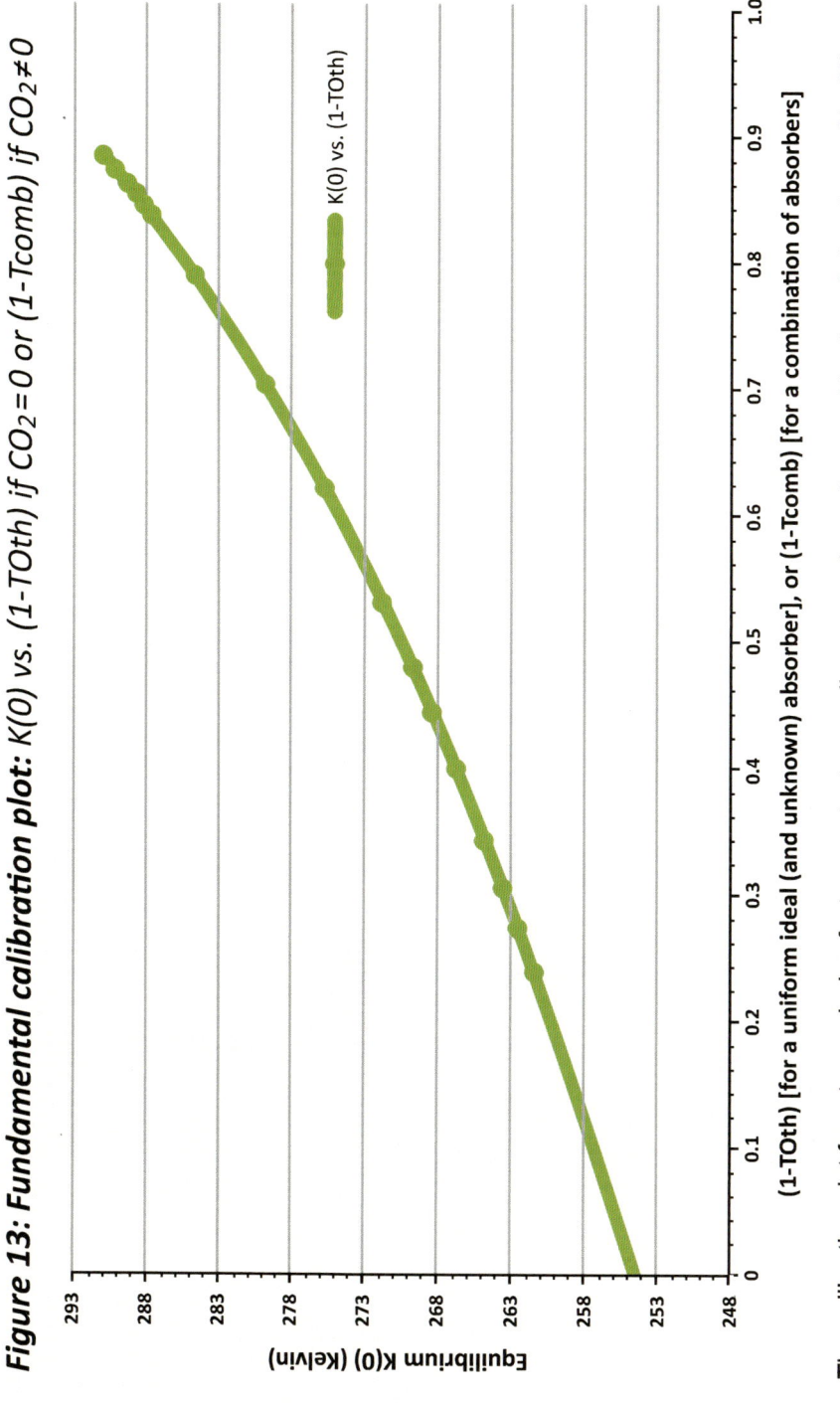

Figure 13: Fundamental calibration plot: $K(0)$ vs. $(1-TOth)$ if $CO_2=0$ or $(1-Tcomb)$ if $CO_2 \neq 0$

The calibration plot for a transmission factor constant over all wavenumbers. Any combination of absorbers with different transmission factors for each WNi has $Tcomb(WNi) = T1(WNi) \cdot T2(WNi) \cdot T3(WNi) \ldots$

$(1-TOth)$ [for a uniform ideal (and unknown) absorber], or $(1-Tcomb)$ [for a combination of absorbers]

Calibration

Figure 14: Highlights of Figure 13 when there is only CO_2 (no background) and when there is CO_2 and a background of TOth = 0.214

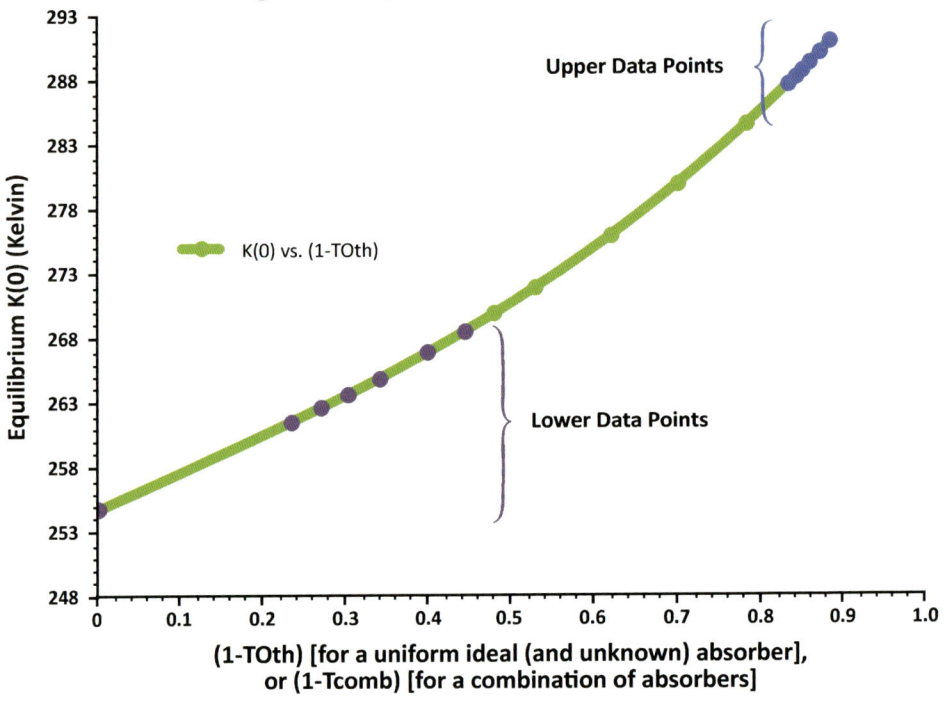

[Plot: Equilibrium K(0) (Kelvin) on y-axis (248 to 293) vs. (1-TOth) [for a uniform ideal (and unknown) absorber], or (1-Tcomb) [for a combination of absorbers] on x-axis (0 to 1.0). Green curve labeled "K(0) vs. (1-TOth)". Upper Data Points clustered near upper right; Lower Data Points clustered in middle-left region.]

The plot in Figure 13 is repeated, as Figure 14, with two areas highlighted. The upper data points have TOth values and K(0) values that exactly match the combination of an unidentified uniform "Other" background with TOth=0.2140 and specified CO_2 values with their own known TCO_2 values for the respective counts of 252.8, 632, 1,264, 2,528, 6,320, and 12,640 ppm.

The "baseline" point is {CO_2=380; K(0)=288.056}; this requires TOther=0.2140. All these data points have TOth values and K(0) values that exactly match the averaged solo TCO_2 values and K(0) values for CO_2 values of 0, 252.8, 632, 1,264, 2,528, 6,320, and 12,640 ppm (0, 200, 500, 1,000, 2,000, 5,000, 10,000 atm-cm) {i.e., calculate K(0) with only CO_2, and then calculate TOther (and (1-TOther)) for each K(0) with CO_2=0.}

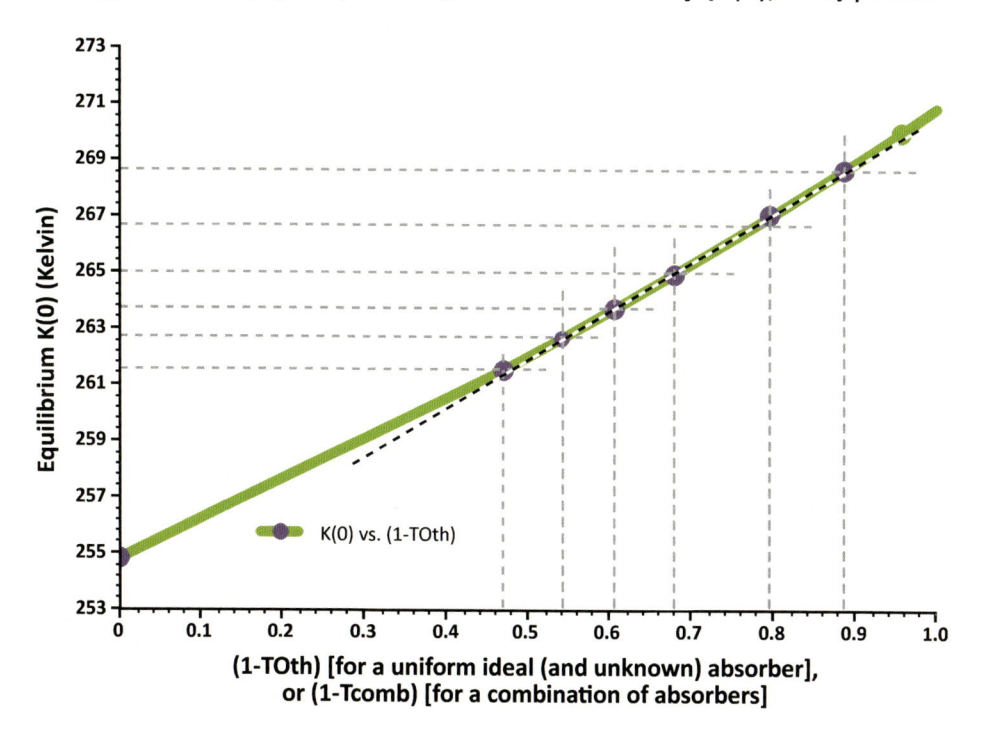

Figure 15: K(0) vs. (1-TOth), with insertions of {K(0), CO₂} points

[Plot: Equilibrium K(0) (Kelvin) on Y-axis from 253 to 273; (1-TOth) [for a uniform ideal (and unknown) absorber], or (1-Tcomb) [for a combination of absorbers] on X-axis from 0 to 1.0; legend: K(0) vs. (1-TOth)]

A magnification of the lower data points in Figure 14. These data points have TOth values and K(0) values that exactly match the averaged solo TCO₂ values and K(0) values for the respective counts of 0, 252.8, 632, 1,264, 2,528, 6,320, and 12,640 ppm {i.e., calculate K(0) with only CO₂, and then calculate TOther for each K(0) with CO₂=0. Such a TOther becomes a suitable effective T(CO₂)}.

Basic calibration plot

One title for **Figure 13** is K(0) vs. (1-TOth) if CO₂=0. With CO₂=0, the only variable is TOth and a simple plot is obtained. The X-axis is chosen as (1-TOth) since (1-TOth) is the effective absorption.

Various values of TOth from 1.0 to 0.1 are selected and K(0)'s are extracted from the spreadsheet.

Actually, this plot addends other points as well. Some points are

Calibration

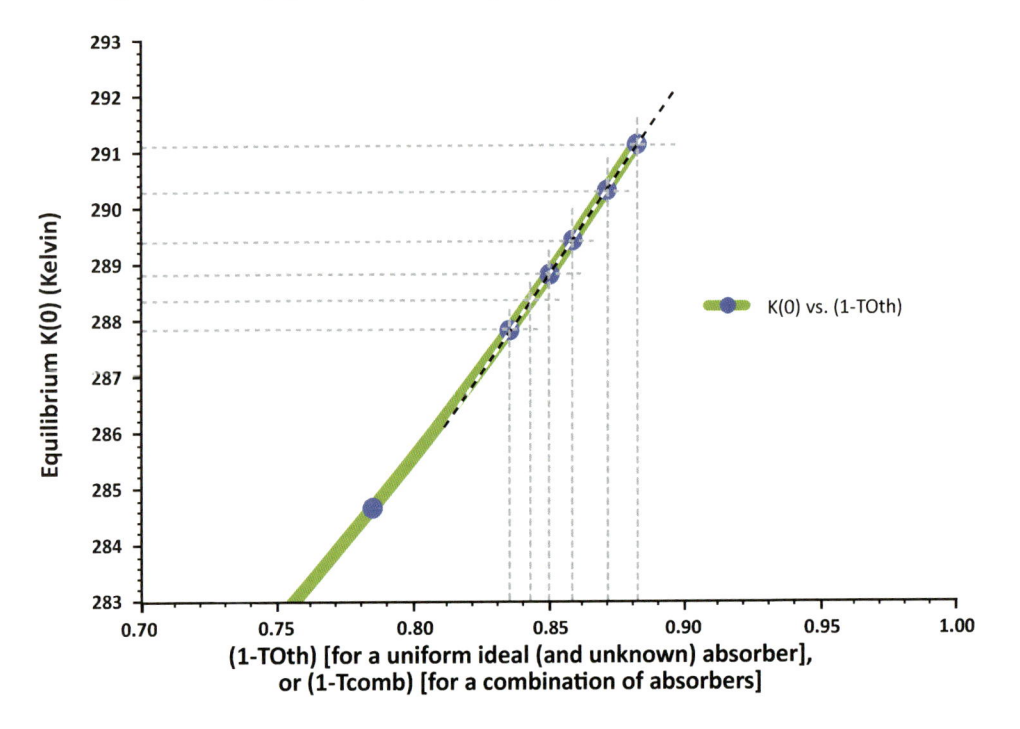

Figure 16: *K(0) vs. (1-Tcomb) or (1-TOth)*

A magnification of the upper data points in Figure 14. The baseline value of K(0) for CO_2=380 ppm is known to be 288.0K. Simulations for CO_2=252.8 and 632 ppm can be interpolated to give the baseline value for CO_2=380 ppm if Tbkgnd=0.2140. The X-axis gives the (1-Tcomb), or equivalent (1-TOth) values If Tbkgnd were to be changed for any or all six purple points (i.e., for all points except for 380 ppm), the purple points would shift, but all would still lay on the calibration plot.

---

for CO_2 absorbers with no background absorber. Even with 12,640 ppm of CO_2, K(0) is never higher than 272K. With CO_2 as the only absorber, doublings of increasing CO_2 will provide ΔK(0) increases of 1.0 to 1.5K.

The K(0) can be transferred onto the calibration plot and

Table 2: *TCO₂ corrected values at lower temperatures*

CO₂ ppm	TCO₂ (low K)	Applicable TCO₂	Stull "ave" TCO₂
0	1	1	1
252.8	0.7645	0.7706	0.7724
380		0.7582	--
632	0.7283	0.73364	0.7013
1,264	0.6964	0.7005	0.671
2,528	0.6591	0.6612	0.6367
6,320	0.6023	0.6	0.5823
12,640	0.5566	0.5491	0.5364

effective TCO₂ values can be assigned. The definition of "effective TOth" is that the K(0) for the stand-alone CO₂ equals the K(0) for a particular value of TOth.

There are additional points with combinations of CO₂ and TOth. One of them is the point:

{K(0)=288.0, CO₂=380, Tcomb
=effTCO₂•TOth=0.7582•0.2140=0.1623}

Determining the effective values of TCO₂

The plot in **Figure 14** is identical to **Figure 13**, with the addition of two areas highlighted with curly brackets.

For the points in the lower left, the spreadsheet first determines the calibration plot with a perfect TOther. The K(0) for CO₂ with no "Other" have been separately calculated via the spreadsheet; these K(0) then provide inputs to the calibration plot and the output is the effective (1-TCO₂).

The effective TCO₂ values for:

are: CO₂={0, 252.8, 632, 1,264, 2,528, 6,320, 12,640} ppm

{1.0, 0.7645, 0.7283, 0.6964, 0.6591, 0.6023, 0.5566}

Calibration

It is interesting to compare the effective TCO_2 values with the algebraic averages of the obviously non-constant $TCO_2(WN)$ values. That sequence gives:

{1.0, 0.7724, 0.7013, 0.6710, 0.6367, 0.5823, 0.5364}

This near-agreement over the full range of CO_2 is quite fortuitous.

The applicable or effective TCO_2 values for applications with a mix of absorbers within the upper data points are given in the column "Applicable TCO_2" of **Table 3**.

K(0) vs. (1-TOth) for modern day K(0)'s

Figure 15 magnifies the lower left quadrant of the calibration plot and **Figure 16** magnifies the upper right of the K(0) vs. (1-TOth) calibration. These K(0) and absorption values are within the range of recent and future K(0).

The spreadsheet calculation for the baseline point of {380 ppm, K(0)=288.0K} requires that there also be a background of TOth=0.214.

The calibration (1-TOth) for K(0)=288.0 is 0.8377. This is also the (1-Tcomb) value for K(0)=288.0 and CO_2=380 ppm. Tcomb=0.1623, and the *eff*TCO_2 value is:

Tcomb/TOth=0.1623/0.214=-0.7582

Continuation of the (1-T0th) & (1-Tcomb) analysis

The previous analysis considered only cases with T0th=0.214, since this was the background value that was essential for (CO_2=380; K(0)=288.0}. Satisfaction of the baseline required that T0th=0.214. This, and extended, results are NOT force-fit to the calibration plot, but fit in a very natural way.

The six plotted points are also calculations from the spreadsheet. For these points, it was only assumed that the background would be the same for them as well, i.e., T0th=0.214. These points also fit perfectly.

The slope near 288.0K(0) is ~+0.67C of K(0) per one one-hundredth (1-T0th), or per one-hundredth of (1-Tcomb).

From the results in the lower left of the calibration plot the increase in K(0) for each doubling of CO_2 when CO_2 is the ONLY absorber can be demonstrated.

From the results in the upper right of the calibration plot the ΔK(0) for each doubling of CO_2 can be calculated. The doubling plot, prepared from first principles with a LOG2(CO_2/380) scale, has already been shown (**Figure 12**).

The only assumption in the latter case for a doubling plot is that T0th is unchanged. However, if increases in T0th for larger CO_2 and decreases for lower CO_2 are specified, new doubling curves can be established.

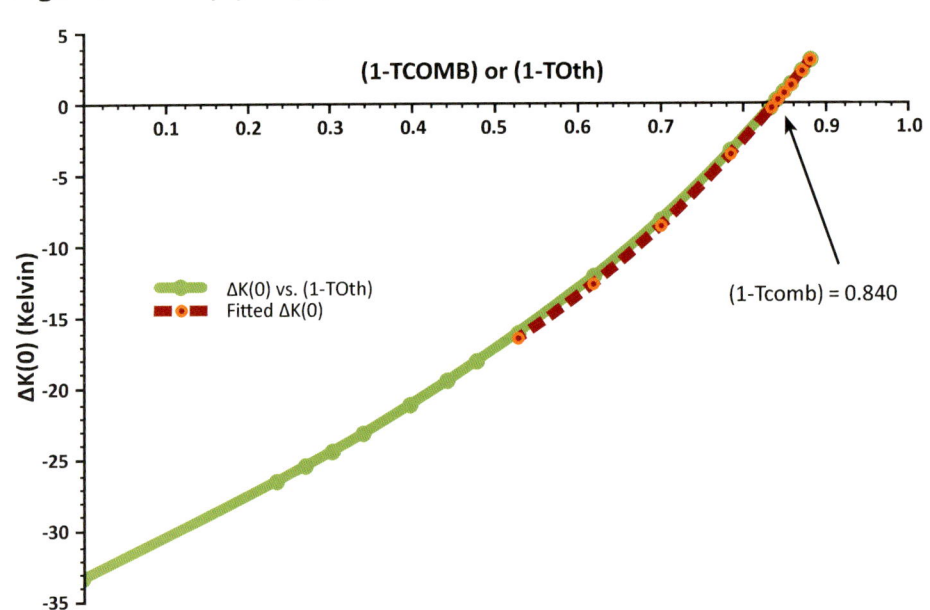

Figure 17: $\Delta K(0) = K(0)-288.056$

When carbon dioxide is the only molecule absorbing energy in the atmosphere, the increase in temperature is far less. Notes: $\Delta K(0)=K(0)-288.056$. The "baseline" point is {CO_2=380; ppm K(0)=288.056K}; this requires TOther=0.2140. The combination (1-Tcomb)=(1-*eff*TCO_2•0.2140) for K(0)=288.0 equals 0.8415 (Figure 16) and so *eff*TCO_2=0.7407.

$\Delta K(0)$ vs. (1-TOth) {or vs. (1-Tcomb)}

The plot in **Figure 17** simply defines $\Delta K(0)$ as K(0)-288.056 and extracts these values from the calibration plot.

It is immediately obvious that the $\Delta K(0)$ values in the "real world" are much lower than when CO_2 is the only absorber. (See earlier "Model Abstract" **Figure 4**.)

When CO_2 is the only absorber, $\Delta K(0)$ when going from 252.8 ppm to 632 ppm is 1.15C (0.87C per doubling); when CO_2 also has a background with TOth=0.214, $\Delta K(0)$ is 0.53C (0.40C per doubling).

$\Delta K(0)$ vs. (1-Tcomb) is the basic approach for evaluating $\Delta K(0)$ for any absorbers. Its overarching value is that it always uses

Global Warming Temperatures and Projections

the same calibration plot, with a CO_2 value associated with each {K(0), (1-Tcomb)} point.

Since the currently published methods favor the presentation of ΔK(0) vs. LOG2(CO_2/380), those plots will also be presented.

A simple (non-optimized) curve fit is also shown. It has a linear term and a quadratic term.

It is clear there is no runaway. *It is also clear that significantly greater counts of absorbing molecules are required in order to achieve equal successive increases of ΔK(0).*

Emphases

- There have been no axes relating to "time." None can be inferred.
- A CO_2 increase from 380 to 760 ppm has a ΔK(0) increase of 0.40 with a Tbkgnd of 0.214 and a ΔK(0) increase of 0.95K with a Tbkgnd of 1.0. Therefore, 0.95K is the maximum doubling value. The calibration plot always remains true no matter what Tbkgnd is. A lower TOth (more absorption) decreases ΔK(0) between any two CO_2 values and decreases the range of (1-Tcomb); a higher TOth (less absorption) increases ΔK(0) and increases the range of (1-Tcomb).
- Therefore, assumptions of (or knowledge of) Tbkgnd for CO_2 greater (or less) than 380 ppm strongly affect the ΔK(0) vs. LOG2(CO_2/380) projections.
- The calibration plot of K(0) vs. (1-Tcomb) is valid for all absorbers. (It does require, however, that valid values for effective TA, TB, etc., be known.)
- The background (almost certainly H_2O) is dominating the present absorption. Stull et al have transmission models for water, but water is not nearly as uniform in either time or altitude as is CO_2. The spreadsheet model, however, will accept an "averaged" water distribution with altitude.
- The Planck distribution for K(Z) always attempts to restore itself. Even 99.9% "notches" for specific WN in the directed BBR at altitude Z and temperature K(Z) are re-filled with non-directed thermalized photons of those WN values, and the filled areas of the notches have no further role in the absorption of

- the remaining BBR. The CO_2 molecules are still present and are still absorbing but the BBR(WN) photon count is already drastically reduced.
- The modeling for post-absorption adiabatic cooling must necessarily result in equal values for the up and down components of thermalized flux. The total post-absorption thermalized flux peaks close to Z' = 0.45 (the maximum horizontal widths of the "orange" pairs) and is zero at Z'=0. The Blanket is the downward-expanding thermalization necessary to increase P(0) to a level that accommodates the daily "escape" of the sun's input. It is an input in the same manner as input from the atmospheric absorption of the sun's input.
- The downward-developing blanket in **Figure 9** is directly matched by the upwards compensation of the (reducing) BBR. If half of the blanket R (red) at any altitude is added to the inner edges of the orange lines (increasing each width of "O" to O+Blkt/2), then the total width, P(Z')=2 • (O+R/2)+(BBR-R)=2 • O+BBR=2 • AdjCumSup+BBR, which is exactly what the total Power should be.
- The "Blanket" as it is calculated in the slice schematic is the mathematical representation of the development of the Blanket from Z'cutoff to Z'=0.
- CO_2 is clearly insignificant compared to Oth and Oth appears to be water vapor. Whereas the effect of CO_2 is only via absorption, it will be shown in Section Two that Nature's water is dominant. It not only absorbs, but cools surfaces upon evaporation, heats the upper atmosphere by means of condensation, and the upper atmosphere conducts this heat by means of convection from the Tropics towards the Poles. The effective T associated with the adiabatic effect is ~0.64. *Eff*TCO_2 is ~0.75; *eff*TH_2O from absorption is ~0.75; *eff*TConv is ~0.5. With a remaining Tbkgnd of 0.9, the final Tcomb product is ~0.162. For the baseline condition of 380 ppm, the component losses in going from ~390W/m² at Z'=0 to 100W/m² at Z'=0.82 are: 140 for adiabatic effect, 20 for BBR escape, 80 for thermal escape, 37.5 for absorption by CO_2, 37.5 for absorption by H_2O, and 75 for the convection blanket plus background. 390=140 +(20+80)+37.5+37.5+75. (Also see **Appendix E**.)

Doubling plots: $\Delta K(0)$ vs. $LOG2(CO_2/380)$

Figure 18 shows the projected temperature increases if CO_2 were the sole absorber, i.e., if there were no background absorption

First of all, with no CO_2, K(0) is projected to be a very cold 255K. If, e.g., the 239 value were reduced to 188 because of increased reflection, P(0) will be 188 and K(0) will be 240K. Increasing the CO_2 to 250 ppm raises the K(0) by almost 7.0C, and a further increase to 630 ppm raises K(0) by another 1.1C.

Icing will, of course, upend this simplistic projection. The act of freezing any weight of water at 0.0C takes 80 times the energy to reduce water temperature by 1.0C. Reflection of the sun's rays by forming ice and snow will also cause a reduction of the 239 W/m² value and a reduction of P(0).

A doubling from 380 ppm to 760 ppm raises K(0) by 0.95C, and doubling again adds another 1.05C. It requires 2,500 ppm (a factor of 6.6) to raise K(0) by 3.0C. CO_2 as a solo absorber clearly has a

Doubling plots

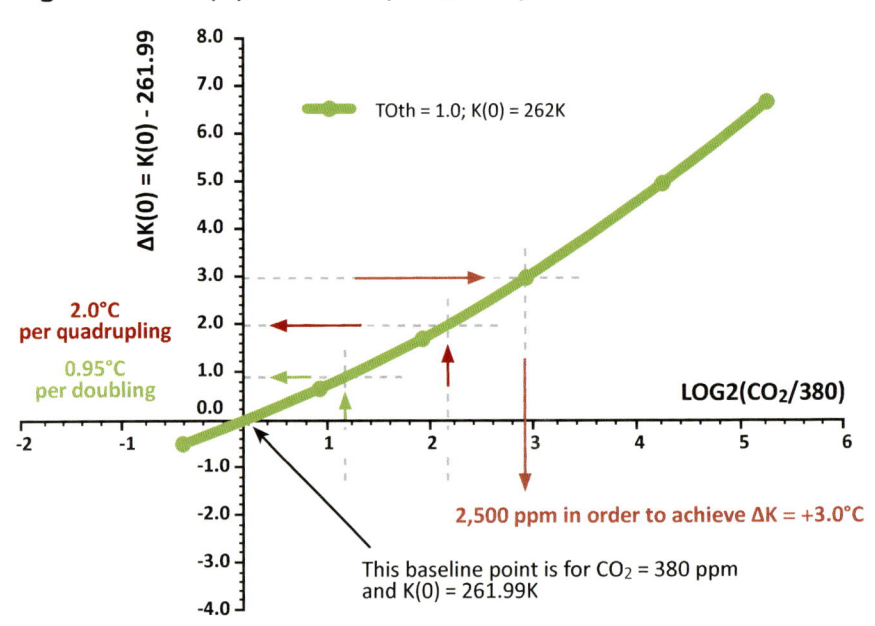

Figure 18: $\Delta K(0)$ vs. LOG2(CO$_2$/380) with TOth = 1.0

[Figure 18: Plot of $\Delta K(0) = K(0) - 261.99$ vs. LOG2(CO$_2$/380) with TOth = 1.0; K(0) = 262K. Annotations: "2.0°C per quadrupling", "0.95°C per doubling", "2,500 ppm in order to achieve $\Delta K = +3.0$°C", "This baseline point is for CO$_2$ = 380 ppm and K(0) = 261.99K".]

Calculations demonstrate that at TOth = 1.0 (modeling a hypothetical situation where carbon dioxide is the only molecule in the atmosphere absorbing energy) doubling the current parts per million of CO$_2$ would raise earth surface temperatures less than one degree Centigrade. Temperatures would rise 3 degrees if the amount of CO$_2$ rose sixfold.

---

very strong effect, but CO$_2$ would have to reach tens of percent in order to obtain any reasonable warming.

If calculations such as this are the arguments behind the CO$_2$ scare, then such conclusions are falsely inferring global calamities.

Figure 19, centered on a baseline point of {380 ppm, 288C}, and a uniform background with TOth=0.214, gives accurate projections. A quadrupling of CO$_2$ will increase K(0) by only 0.9C. 4,800 ppm (a factor of 12!) is required to increase K(0) by +2.0C.

Global Warming Temperatures and Projections

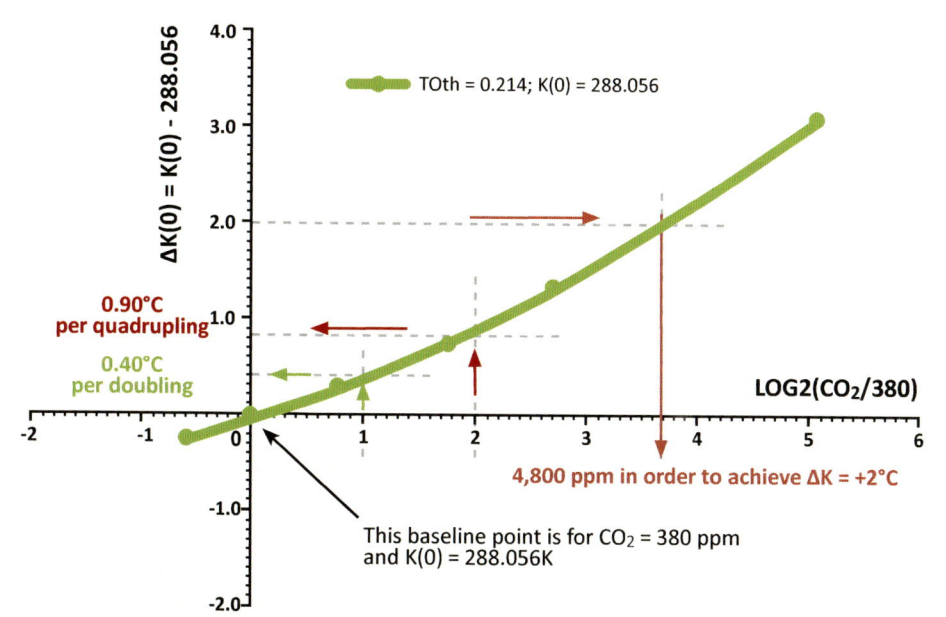

Figure 19: $\Delta K(0)$ vs. $LOG2(CO_2/380)$ with $TOth = 0.214$

[Figure 19: Graph of $\Delta K(0) = K(0) - 288.056$ vs. $LOG2(CO_2/380)$, with curve labeled $TOth = 0.214$; $K(0) = 288.056$. Annotations: "0.90°C per quadrupling", "0.40°C per doubling", "This baseline point is for $CO_2 = 380$ ppm and $K(0) = 288.056K$", "4,800 ppm in order to achieve $\Delta K = +2°C$".]

In the atmosphere as currently constituted, a four-fold increase in CO_2 would raise the earth's surface temperature by only 0.9C. If the CO_2 molecules reached 4,800 parts per million -- twelve times the existing level -- the surface temperature would increase by 2.0 degrees C. Figure 19 is a duplicate of Figure 12.

Assessing the assumption that TOth is constant as CO_2 increases

All of the previous calculations show that CO_2 is not a villain. It is not even a "pollutant" of any kind. It is abundantly important to our lives.

The only way in which CO_2 can be a cause of genuine pollutants is when the burning of carbon fuels introduces particulates into the atmosphere. Such particulates do not cause perpetually increasing temperatures either, but do cause the unhealthy air of cities such

Doubling plots

Figure 20: *Adjusting TOth to modify K(0) increases*

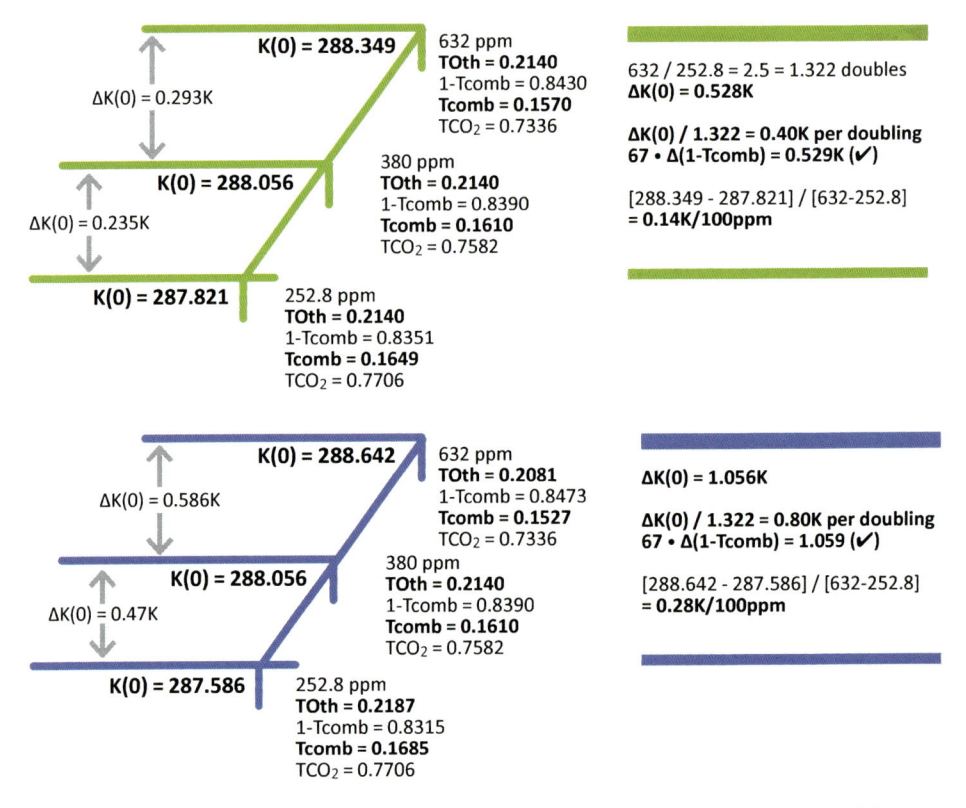

Upper (green) progression:

- K(0) = 288.349 ← 632 ppm, TOth = 0.2140, 1-Tcomb = 0.8430, Tcomb = 0.1570, TCO_2 = 0.7336
- ΔK(0) = 0.293K
- K(0) = 288.056 ← 380 ppm, TOth = 0.2140, 1-Tcomb = 0.8390, Tcomb = 0.1610, TCO_2 = 0.7582
- ΔK(0) = 0.235K
- K(0) = 287.821 ← 252.8 ppm, TOth = 0.2140, 1-Tcomb = 0.8351, Tcomb = 0.1649, TCO_2 = 0.7706

632 / 252.8 = 2.5 = 1.322 doubles
ΔK(0) = 0.528K

ΔK(0) / 1.322 = 0.40K per doubling
67 • Δ(1-Tcomb) = 0.529K (✔)

[288.349 - 287.821] / [632-252.8]
= 0.14K/100ppm

Lower (blue) progression:

- K(0) = 288.642 ← 632 ppm, TOth = 0.2081, 1-Tcomb = 0.8473, Tcomb = 0.1527, TCO_2 = 0.7336
- ΔK(0) = 0.586K
- K(0) = 288.056 ← 380 ppm, TOth = 0.2140, 1-Tcomb = 0.8390, Tcomb = 0.1610, TCO_2 = 0.7582
- ΔK(0) = 0.47K
- K(0) = 287.586 ← 252.8 ppm, TOth = 0.2187, 1-Tcomb = 0.8315, Tcomb = 0.1685, TCO_2 = 0.7706

ΔK(0) = 1.056K

ΔK(0) / 1.322 = 0.80K per doubling
67 • Δ(1-Tcomb) = 1.059K (✔)

[288.642 - 287.586] / [632-252.8]
= 0.28K/100ppm

This figure shows a pair of progressions as CO_2 goes from 252.8 ppm to 380 ppm to 632 ppm. The upper sequence has a fixed TOth = 0.214. A doubling of CO_2 produces a ΔK(0) of 0.40C per doubling of CO_2 and an increase of 0.14C per 100 ppm. The lower progression increases TOth to 0.2187 for CO_2 = 252.8 and decreases TOth to 0.2081 for 632 ppm. Both ΔK(0)'s for the first lower progression are doubled despite TOth changes of only ±2.5%

as Beijing. (Rain scrubs the air of these particulates and so they do not continue to build up, a fortunate — but insufficiently satisfying — result.)

Water is the dominant absorber in the atmosphere and it also could unjustifiably be called a pollutant that increases temperatures.

Global Warming Temperatures and Projections

Table 3: Extended (gedanken) table from 6 ppm to 12,640 ppm

CO_2 ppm Input	LOG2 (CO_2/380)	Stepped TOth	Proj & Known TCO_2	(1-Tcomb)	Add'l effect from CO_2 (1-Tcomb) -(1-TOth)	% add'l eff from CO_2	K(0)
5.9375	-6	0.25	0.8974	0.7757	0.0257	3.31%	284
11.875	-5	0.244	0.8759	0.7863	0.0303	3.85%	284.7
23.75	-4	0.238	0.8524	0.7971	0.0351	4.40%	285.4
47.5	-3	0.232	0.8279	0.8079	0.0399	4.94%	286.1
95	-2	0.226	0.8024	0.8187	0.0447	5.46%	286.75
190	-1	0.22	0.7759	0.8293	0.0493	5.94%	287.45
252.8	-0.588	0.2175	0.7706	0.8324	0.0499	5.99%	287.645
380	0	0.214	0.7582	0.8377	0.0517	6.17%	288
632	0.7339	0.2096	0.73364	0.8462	0.0558	6.59%	288.567
1,264	1.7339	0.2036	0.7005	0.8574	0.0610	7.11%	289.335
2,528	2.7339	0.1976	0.6612	0.8693	0.0669	7.70%	290.179
6,320	4.0559	0.1897	0.6	0.8862	0.0759	8.56%	291.428
12,640	5.0559	0.1837	0.5491	0.8991	0.0828	9.21%	292.432

Or as a "pollutant" because it contributes to hurricanes and floods.

Even as UN scientists continue to "adjust" actual temperature records, their models continue to overpredict temperature increases over time. The author's calibrated model, however, underpredicts the official temperature increases when compared to officially measured $\Delta K(0)$ per 100 ppm increases in CO_2.

When official data is used, the apparent increase in temperature as CO_2 is increased is nonlinear but, over a narrow range, is about 0.25C per 100 ppm. The author's model with constant TOth=0.214

Doubling plots

Figure 21: Possible TCO$_2$ extension to CO$_2$ values less than 252.8 ppm

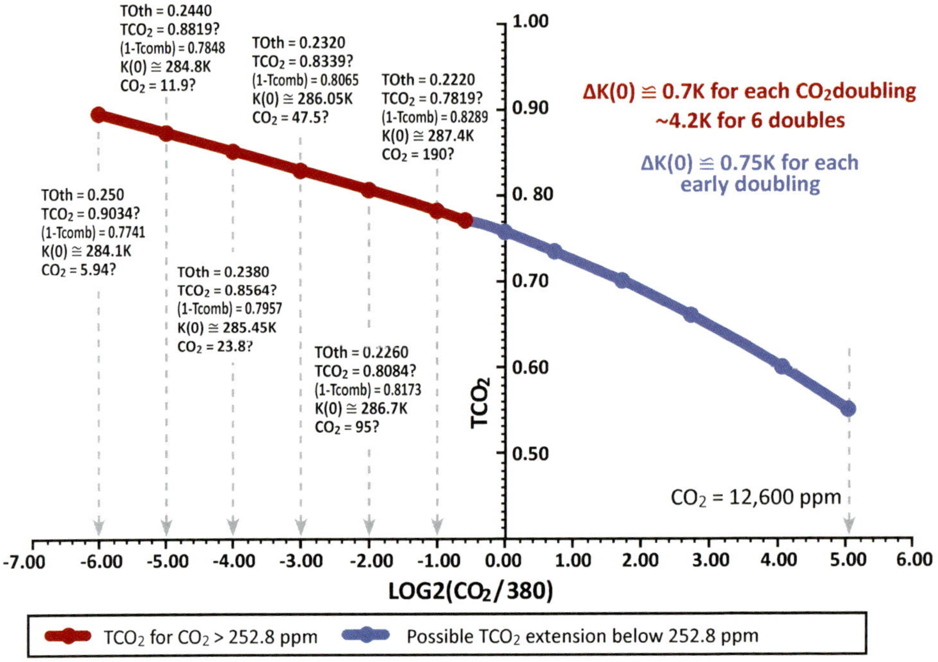

TOth = 0.2440
TCO$_2$ = 0.8819?
(1-Tcomb) = 0.7848
K(0) ≅ 284.8K
CO$_2$ = 11.9?

TOth = 0.2320
TCO$_2$ = 0.8339?
(1-Tcomb) = 0.8065
K(0) ≅ 286.05K
CO$_2$ = 47.5?

TOth = 0.2220
TCO$_2$ = 0.7819?
(1-Tcomb) = 0.8289
K(0) ≅ 287.4K
CO$_2$ = 190?

ΔK(0) ≅ 0.7K for each CO$_2$ doubling
~4.2K for 6 doubles

ΔK(0) ≅ 0.75K for each early doubling

TOth = 0.250
TCO$_2$ = 0.9034?
(1-Tcomb) = 0.7741
K(0) ≅ 284.1K
CO$_2$ = 5.94?

TOth = 0.2380
TCO$_2$ = 0.8564?
(1-Tcomb) = 0.7957
K(0) ≅ 285.45K
CO$_2$ = 23.8?

TOth = 0.2260
TCO$_2$ = 0.8084?
(1-Tcomb) = 0.8173
K(0) ≅ 286.7K
CO$_2$ = 95?

CO$_2$ = 12,600 ppm

LOG2(CO$_2$/380)

— TCO$_2$ for CO$_2$ > 252.8 ppm   — Possible TCO$_2$ extension below 252.8 ppm

A plot illustrating the extrapolation of CO$_2$ values presented in Table 3.

---

suggests values of about 0.14C per 100 ppm (over the 379.2 range from 252.8 ppm to 632 ppm).

It was the baseline requirement that K(0) should be equal to 288K for a CO$_2$ level of 380 ppm. This requirement leads to the corollary requirement that all other possible absorbers have an effective transmission value TOth=0.214 for CO$_2$ equal to 380 ppm. There is no second baseline requirement that either concludes or negates whether this TOth applies to other CO$_2$ values.

However, the calibration plot provides a way of estimating what TOth (i.e., Tbgrnd) must be for other CO$_2$ values if one chooses to have faith in the accuracy of the officially observed ΔK(0) per doubling. That is, the calibration plot, in alliance with the spreadsheet program, can operate in either direction.

Figure 20 gives sketches of the transition from 0.40C to 0.80C per doubling of CO_2 as TOth is adjusted AND to a corresponding doubling of $\Delta K(0)$ per 100 ppm CO_2. The background is more important than the CO_2.

Table 3 extrapolates from the TOth sequence in **Figure 20**. The value of TOth=0.2140 is retained for 380 ppm, but is decreased for larger values of CO_2 (more absorption by the background) and is increased for lower values of CO_2 (less background absorption).

There is nothing "scientific" about these extrapolations; they are simply projections from the result which doubled the $\Delta K(0)$ rate per 100 ppm of CO_2 on either side of 380 ppm. They do, however, demonstrate that other extrapolations can also be easily modeled.

Some results are also given in this table. These results have been extracted from subsequent supporting figures.

Figure 21 gives a graphic interpretation of the early columns in **Table 3**.

Comparisons with the calibration plot

Figure 22 shows K(0) vs. (1-Tcomb) with CO_2 from 6 to 12,640 ppm and with corresponding TOth stepped from 0.250 to 0.1837. (See **Table 3**.)

The same figure shows the calibration curve containing only TOth from 0.250 to 0.1837 (i.e., with CO_2=0).

The plot of K(0) with CO_2 vs. (1-Tcomb) overlays nearly perfectly onto the calibration plot of K(0) vs. (1-TOth)

The previous doubling results obtained via LOG2(CO_2/380) and the linear results obtained via Δ(1-Tcomb) are confirmed.

Figure 23 shows K(0) for the above example plotted vs. LOG2(CO_2/380) over the full range of CO_2 from 6 ppm to 12,640 ppm.

ΔK(0)≅0.7K for each CO_2 doubling for CO_2<250 ppm (~4.2K for 6 doubles).

These extensions are just examples. In reality the TOth values may decrease more significantly; the doubling value for ΔK(0) will then increase.

ΔK(0) ≅0.77K for each early doubling above CO_2=380 ppm.

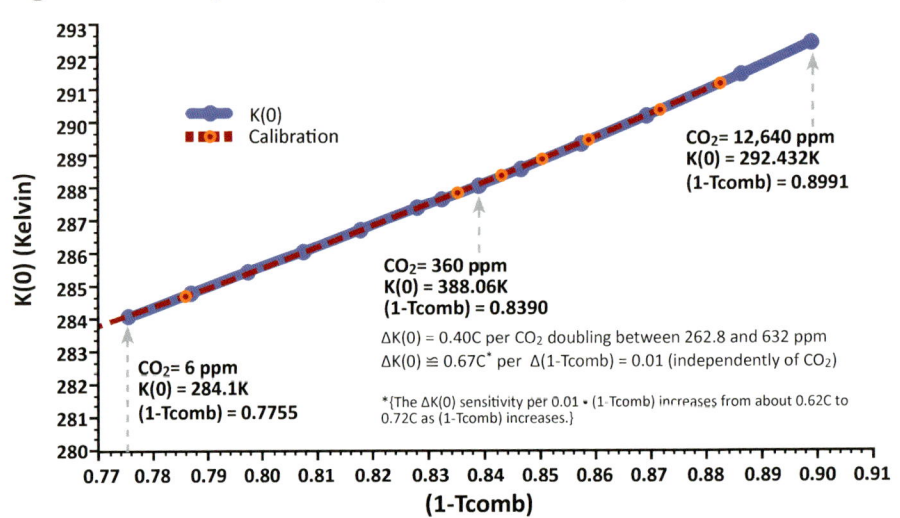

Figure 22: Confirmation of the calibration plot

K(0) vs. (1-Tcomb) **with CO_2 from 6 to 12,640 ppm and with TOth from 0.250 to 0.1837** along with the **calibration containing only TOth from 0.250 to 0.1837.** The overlap validates the calibration models.

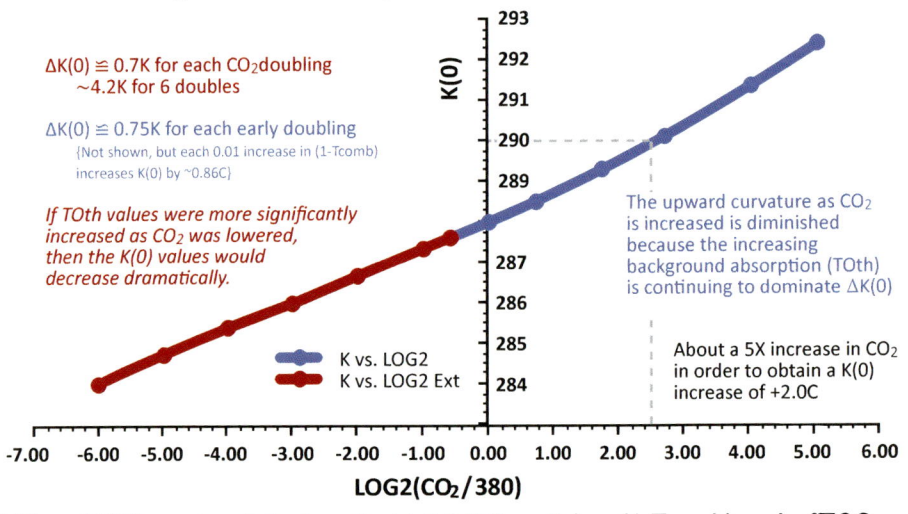

Figure 23: Doubling plot with the same data as Table 3, Figure 21 and Figure 22

K(0) or ΔK(0) can be plotted against LOG(CO_2 ratio) or (1-Tcomb) and *eff*TCO_2 can be plotted against LOG(CO_2 ratio). This plot shows K(0) vs. $LOG_2(CO_2/380)$ with modified extrapolations of TOth from its value of 0.2140 for 380 ppm of CO_2.

Global Warming Temperatures and Projections

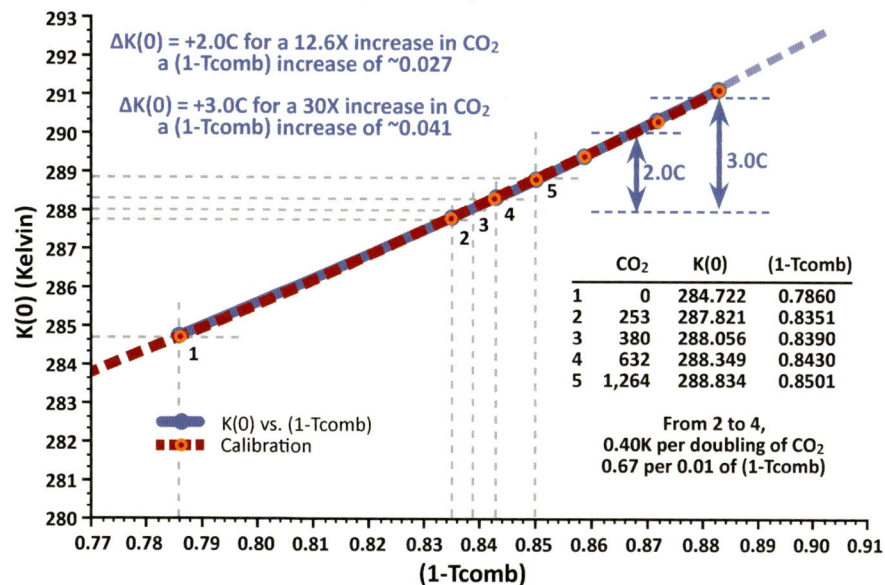

Figure 24: Mix of varying TOth, fixed TOth and CO_2

ΔK(0) = +2.0C for a 12.6X increase in CO_2
a (1-Tcomb) increase of ~0.027

ΔK(0) = +3.0C for a 30X increase in CO_2
a (1-Tcomb) increase of ~0.041

	CO_2	K(0)	(1-Tcomb)
1	0	284.722	0.7860
2	253	287.821	0.8351
3	380	288.056	0.8390
4	632	288.349	0.8430
5	1,264	288.834	0.8501

From 2 to 4,
0.40K per doubling of CO_2
0.67 per 0.01 of (1-Tcomb)

Figure 24 is a replay and extension of Figure 16, with specific examples of how such a plot can be interpreted. K(0) vs. (1-Tcomb) with **TOth=0.214 and CO_2 from 0.0 to 12,640 ppm** along with the **calibration containing only TOth from 0.2987 to 0.1175.**

How to read the calibration plot

Figure 24 repeats the previous comparisons when TOth is left fixed, equal to 0.2140, as CO_2 is changed. The overlap with the calibration plot is excellent. It provides specific examples for interpreting the plot.

A +2.0C increase in K(0) requires CO_2 to be increased by twelve times. A +3.0C increase in K(0) requires CO_2 to be increased by thirty times.

Comparisons with the calibration plot

Figure 25: *P(0)*, *Blkt(0)* and *(1-Tcomb)* vs. *K(0)* Calibrations

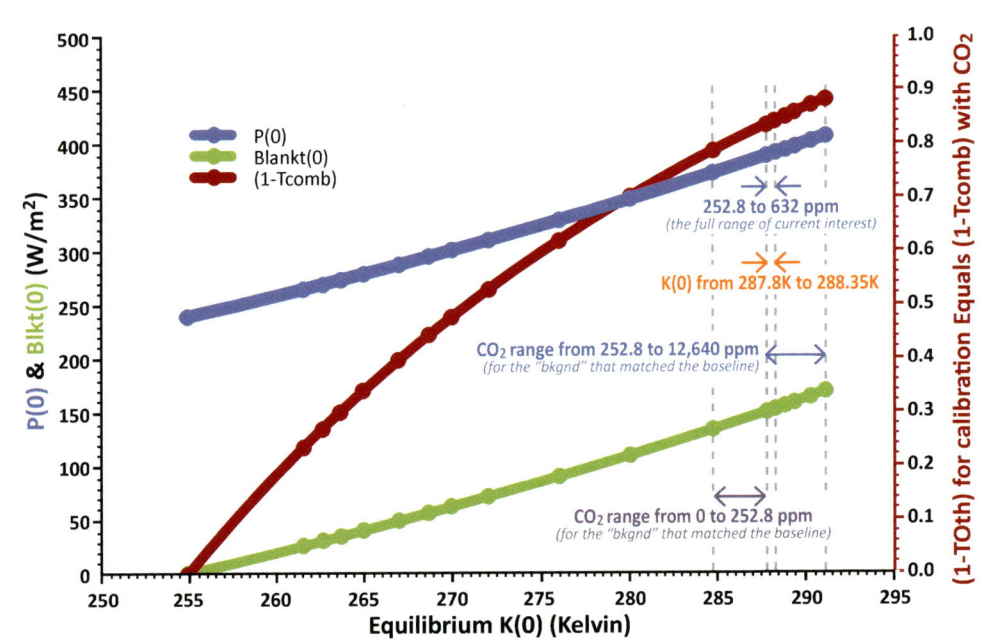

[Figure 25: Composite calibration plot showing P(0), Blkt(0), and (1-Tcomb) plotted against Equilibrium K(0) in Kelvin. Left axis: P(0) & Blkt(0) (W/m²) ranging 0-500. Right axis: (1-TOth) for calibration Equals (1-Tcomb) with CO_2, ranging 0.0-1.0. X-axis: Equilibrium K(0) (Kelvin) from 250 to 295. Annotations: "252.8 to 632 ppm (the full range of current interest)", "K(0) from 287.8K to 288.35K", "CO_2 range from 252.8 to 12,640 ppm (for the 'bkgnd' that matched the baseline)", "CO_2 range from 0 to 252.8 ppm (for the 'bkgnd' that matched the baseline)".]

Composite calibration plots of P(0), Blkt(0) and (1-Tcomb) vs. K(0). All equilibrium points lay directly on the (1-TOth) calibration plot (where an ideal absorber is the only absorber). The K(0) equilibrium values for any set of (1-Tcomb), no matter what the CO_2 (e.g.) ppm and background absorbers might be, fall on this calibration plot. The "CO_2 ranges" as shown here depend, however, on how strong those "Other" absorbers are.

Composite calibration plots (Figures 25 and 26)

When, e.g., CO_2 is fixed at 2,528 ppm (a factor of 10 over 252.8) and TOth is allowed to range from 1.0 to 0.1 (a range of 10), the range for ΔK(0) is 30K! That's 9.0K per doubling, but it is a doubling of (1-TOth). CO_2 is playing no role. The correlation with the basic calibration plot is still accurate.

In summary, all variations of TOth (or Tcomb) and CO_2 validate the basic calibration plot.

Global Warming Temperatures and Projections

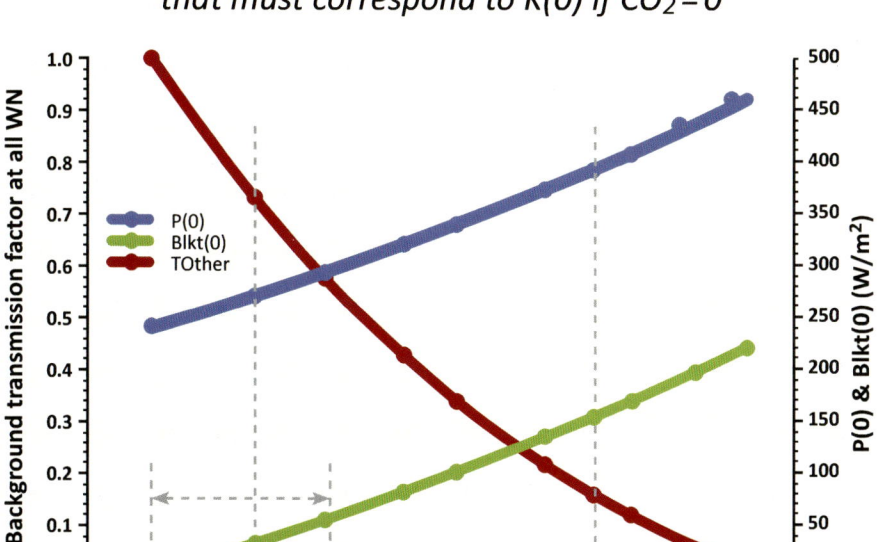

Figure 26: TOther, P(0), and Blkt(0) that must correspond to K(0) if $CO_2 = 0$

Composite calibration plots of P(0), Blkt(0) and Tcomb vs. K(0).

Notes:

The red curve describing the solo TOther (when $CO_2=0$) is also the effective combined TComb for any equilibrium K(0).

The purple curve P(0) and the green curve Blkt(0) are totally dictated by K(0), no matter what the absorbers are.

The vertical line at K(0)=288.349K is the stable K(0) when $CO_2=632$ ppm, but it requires a solo bkgnd TOther of 0.2140. If there were no CO_2, TOther would have to be 0.1568.

CO_2 of 632 ppm, with no other absorbers, has a K(0) of only 262.66K. Even a CO_2=12,640 ppm has a K(0) of only 268.04K. The effective T of CO_2 at 288.349K is 0.7327, since (0.7327)(0.2140)=0.1568. A solo bkgnd absorption is about 3 times greater than the additional CO_2 absorption.

Additional details on absorption

A previous chapter was titled as "A Very Brief Summary of the Spreadsheet Model." Eight previous figures provided a glimpse of "calculations within each slice." **Appendix C** gives more details.

The following **Figure 27** is an updated "refresh" of the slice discussion. Subsequent figures provide additional analysis of the absorption.

As already discussed, the schematic presents the full-fledged model for fluxes associated with a single $\Delta Z'$ slice. To the left is lower Z'; to the right is higher Z'. As the BBR flux progresses upward some is lost to adiabatic expansion and some is absorbed. However, no energy is lost by an absorption; the only aspect that may be lost is its directionality. The photon post-absorption emissions are isotropic (non-directional) and are no longer considered in discussions of absorptions. Energy loss occurs only as thermodynamically defined adiabatic work/loss or as true "escape."

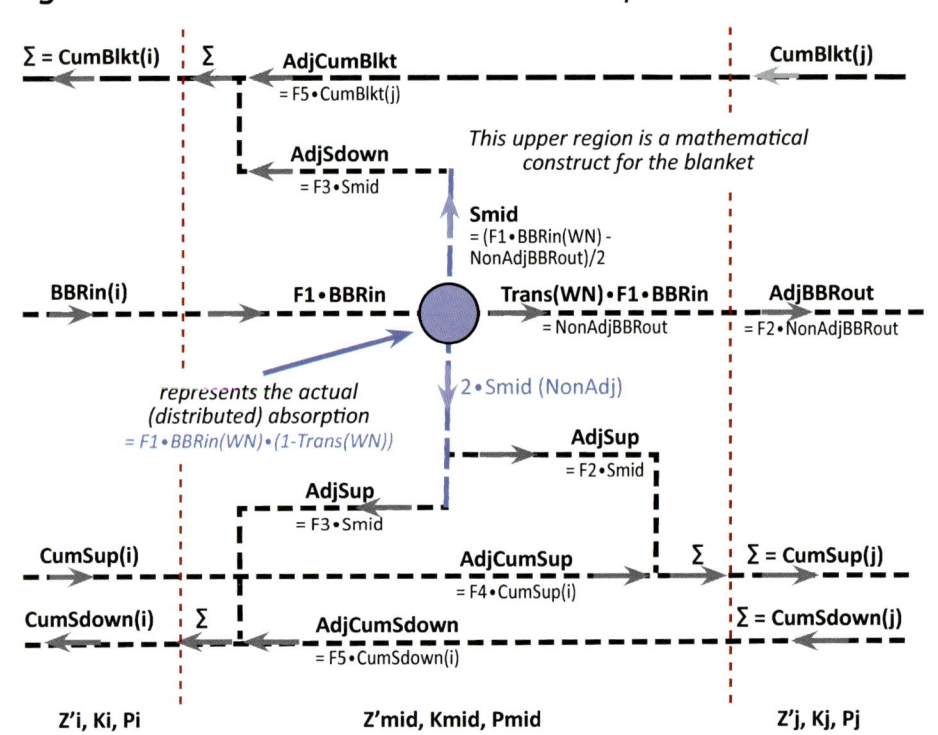

Figure 27: Interactions within each atmospheric slice

Top row (left to right): Σ = CumBlkt(i) ← Σ ← AdjCumBlkt = F5•CumBlkt(j) ← CumBlkt(j)

Middle upper: AdjSdown = F3•Smid

"This upper region is a mathematical construct for the blanket"

Smid = (F1•BBRin(WN) − NonAdjBBRout)/2

Middle row: BBRin(i) → F1•BBRin → [circle] → Trans(WN)•F1•BBRin = NonAdjBBRout → AdjBBRout = F2•NonAdjBBRout

"represents the actual (distributed) absorption = F1•BBRin(WN)•(1−Trans(WN))"

2•Smid (NonAdj)

AdjSup = F2•Smid

AdjSup = F3•Smid

Lower rows:
CumSup(i) → → AdjCumSup = F4•CumSup(i) → Σ → Σ = CumSup(j)

CumSdown(i) ← Σ ← AdjCumSdown = F5•CumSdown(i) ← Σ = CumSdown(j)

Column labels: Z'i, Ki, Pi | Z'mid, Kmid, Pmid | Z'j, Kj, Pj

A schematic of the interactions to be tracked within each slice: BBR absorption, development of thermalized (up and down) radiation, and cumulative downwards growth of the "Blanket." The CumBlkt terms are summed from Z' = Z'cutoff to Z' = 0. The CumSdown terms show left-pointing arrows, but they are summed from Z' = 0. {See Appendix C for an improved explanation.}

Even with no atmospheric infrared absorbers, a sun's daily input of 239 W/m² (with no reflective return, the net input would be higher than this) would still be dissipated daily by adiabatic expansion.

With absorption, and with the surface at an equilibrium of 288.0K, the adiabatic work/loss at Z' = 0.82 (Z = 13.13 km) is about 139 W/m². The ratio of 139 to P(0) is about 0.36. For equilibrium cases, the adiabatic loss at Z' = 0.82 is always approximately 139.

The upper half of the schematic concentrates on the mathematical

Global Warming Temperatures and Projections

Table 4: Summary table with TOth = 0.2140

CO_2 (ppm)	K(0) (Kelvin)	P(0) (W/m²)	CumAbs (Z'=0.82)	Blkt (0) (W/m²)	Blkt(0) / CumAbs
0	284.722	372.621	199.101	133.62	0.671
632	288.349	391.974	236.832	152.976	0.646
1,264	288.834	394.618	241.933	155.619	0.643
6,320	290.353	402.985	258.434	163.986	0.635

construct of the Blanket development. What was a single track at the bottom is now a double track representing the total thermalization. Their combination is the 2•AdjCumS, with equal upwards and downwards radiation, which has been presented earlier in **Figure 6**.

The down path, AdjCumBlkt, is difficult to calculate directly since its summation begins far off to the right at Z'cutoff. AdjCumBlkt is the (mathematical) accumulation of downward-directed thermalized radiation that eventually produces Blkt(0). All of AdjCumBlkt is produced post-absorption.

The mathematical construct of CumBlkt is not a "flow" independent of AdjCumS. It is a calculation that defines what K(0) must be (for a fixed CO_2 and TOth) in order that the required Blkt(0) is obtained. It is that K(0) and the absorption that then determine the correct AdjCumS. The Blanket is thermalized (not a BBR), and is an additive component to the "orange" component of thermalization in **Figures 9, 10, and 11**.

AdjCumSup and AdjCumSdown are identical at all Z'. Their summations both begin at Z'=0 and are necessarily equal since absorptions produce post-absorption emissions that are isotropic, equally probable for upwards or downwards directivity. The adiabatic effect, the reduction with altitude, affects the composite thermalized flux, which itself has no directivity.

If there were no absorption, along with no input from the sun, the sustainability of a steady state emission from the surface would require a steady state input equal to P(K(0)). But we do have the sun

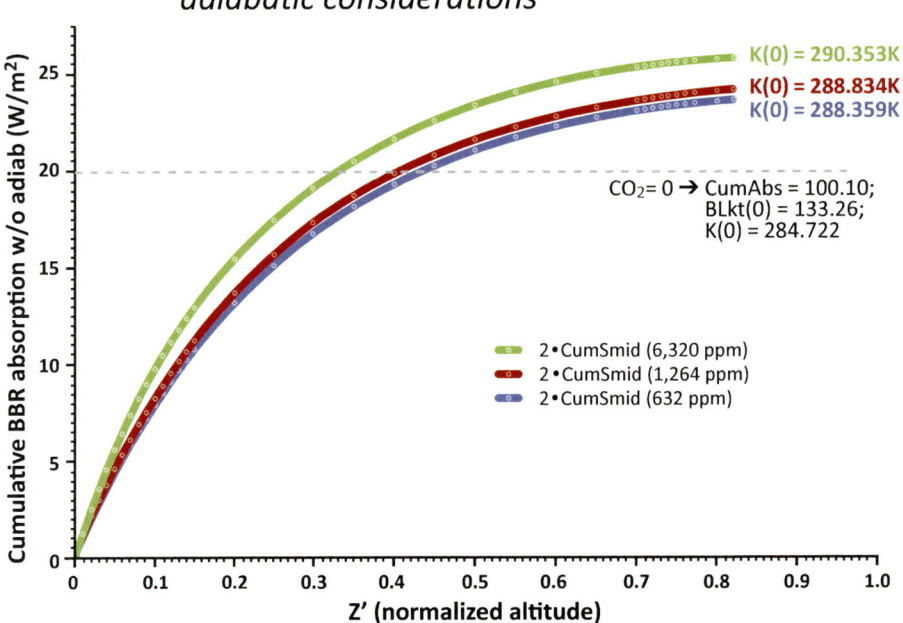

Figure 28: *Cum BBR direct absorption without adiabatic considerations*

Curve labels:
- K(0) = 290.353K
- K(0) = 288.834K
- K(0) = 288.359K

CO_2 = 0 → CumAbs = 100.10; Blkt(0) = 133.26; K(0) = 284.722

- 2•CumSmid (6,320 ppm)
- 2•CumSmid (1,264 ppm)
- 2•CumSmid (632 ppm)

Y-axis: Cumulative BBR absorption w/o adiab (W/m²)
X-axis: Z' (normalized altitude)

The blue curve describes **632 ppm**, **K(0)=288.349K**; TOth=0.2140
- **Cumulative BBR Absorption=Sum of 2 • Smid for each slice=236.83.**
- **P(0)=391.97**; P(0.82)=97.23 ; BBR(0.82)=20.00 ; AdjCumSup(0.82)=37.93.
- [P(0)-P(0.82)]-BBR(0.82)-AdjCumSup(0.82)=236.81
- **Blkt(0)=152.98**; **Blkt(0)/CumBBR Abs=0.646**

The red curve describes **1,264 ppm**, **K(0)=288.834K**; TOth=0.2140
- **Cumulative BBR Absorption=Sum of 2 • Smid for each slice=241.93.**
- **P(0)=394.62**; P(0.82)=97.23 ; BBR(0.82)=19.41 ; 2 • AdjCumSup(0.82)=76.64.
- [P(0)-P(0.82)]-BBR(0.82)-AdjCumSup(0.82)=239.75
- **Blkt(0)=155.62 ; Blkt(0)/CumBBRAbs=0.643**
- [(P(0, 1,264)-P(0, 632)]/[(CumAbs(0.82, 1,264)-CumAbs(0.82, 632)]=0.52
- [(Abs(0.82, 1,264)-Abs(0.82, 632)] /[(K(0, 1,264)-K(0, 632)]=5.45 W/m² /K

The green curve describes **6,320 ppm**, **K(0)=290.353K**; TOth=0.2140
- **Cumulative BBR Absorption=Sum of 2 • Smid for each slice=258.43.**
- **P(0)=402.98** ; P(0.82)=97.23 ; BBR(0.82)=17.72 ; AdjCumSup(0.82)=39.55.
- [P(0)-P(0.82)]-BBR(0.82)-AdjCumSup(0.82)=248.48
- **Blkt(0)=163.986 ; Blkt(0)/CumBBR Abs=0.635**
- [(P(0, 6,320)-P(0, 1,264)]/[(CumAbs(0.82, 6,320)-CumAbs(0.82, 1,264)]=0.51
- [(Abs(0.82, 6,320)-Abs(0, 1,264)]/[(K(0, 1,264)-K(0, 632)]=5.51 W/m² per 1.0K

and we do have absorption. The absorption is creating up and down post-absorption emissions that wouldn't otherwise exist. The BBR surface emission associated only with the sun is no longer sufficient. The blanket, which occurs only because of the absorption, provides additional surface priming. This increases the surface BBR, but also the net absorption and the new Blanket. This continues, but not in a runaway fashion. A surface temperature is reached that satisfies the need to daily emit each day's input from the sun as well as from the absorptions.

Whatever is producing the increased absorption is not likely to also increase the net input of 239 W/m^2 from the sun. If the sun's input is retained, and there is 100% absorption of the surface BBR, and TOth were equal to 0.05 (three times worse than the present Tcomb), then $\Delta K(0)$ would be 8.8C. This is not a reasonable forecast. If, e.g., increased reflection reduced the sun's input to 230 W/m^2 then the ΔK of 8.8C would be reduced by about 2.0C. A ΔK of 8.8C could not be justified by, and it definitely could not be caused by, increased CO$_2$ and some fantastic multiplication of a water effect. (The author has a detailed (unpublished) document on the factors involved in increasing K(0).)

Cumulative sum of BBR absorptions

The thought is simple; the bookkeeping is tricky. The chosen granularity for a slice is sufficient to maintain valid boundary conditions since no disturbing offsets appear in the full plots. (Also see **Appendix E.**)

Figure 28 shows the cumulated sum (summing across WN bands, plus summing from slice to slice) of the actual absorptions in each slice. Although the remanent BBR (WN) entering each slice has experienced pre-absorption adiabatic loss, it continues to experience adiabatic loss within the slice up until each absorption event.

The remanent BBR(WN) after absorption also experiences adiabatic loss from, effectively, {Z'mid, Plmid} to {Z'up, Plup}. This BBR(WN) input to the next slice also experiences an adiabatic loss from {Z'down, Pldown} to {Z'mid, Plmid} of the next slice, and this identifies it as the average pre-absorption BBR(WN) in that slice.

More discussion will follow, but a summary TOth=0.2140 for all CO_2 is provided in **Table 4**. Bolded data values below **Figure 28** correspond with entries in **Table 4**.

Figure 28 shows plots of the accumulated sum of BBR absorptions with altitude. The plot is necessarily sublinear since there is a lower remanent BBR input to each slice.

These are the absorptions that create both the warming downward "blanket," via Sdown, and the upward emissions which can only be eliminated by the adiabatic effect. Nature accommodates by automatically raising the temperature gradients so that the cooling adiabatic loss rate is increased.

The ratios of CO_2 for subsequent plots are 2.0 and 5.0, for a total ratio of 10. These ratios are much greater than the much lower ratios that lately are of great concern in the media, but there is nothing disconcerting about the results. A ten-fold increase in CO_2 is NOT increasing total absorption by 1,000 percent, but only by 9 percent. A reasonable person should not be alarmed.

The summaries are also interesting. Each 1.0 increase in K(0), or about 5.5 W/m² increase in P(0), requires an increase in absorption of about 11 W/m². This is nature's safety factor of two against any possible runaway, something that is very desirable for highway and rail bridges, but is not a problem here.

The ratio of Blanket at Z'=0 to CumAbs at Z'=0.82 (the chosen "cutoff" for adiabatic effects) consistently shows values of about 0.64. All is calm.

Appendix A provides some additional correlations with total absorption.

How much additional absorption does CO_2 provide?

This analysis has been concentrating on the direct effect CO_2 has on global temperature. One conclusion is that TOth is providing most of the absorption (whether or not CO_2 is a direct or indirect agent of any increases of TOth).

When there are two distinct absorbers, the total absorption is NOT the SUM of what each would produce by itself. The addition of a second absorber—in this case, CO_2—provides an additional absorption to what the main absorber—in this case, "Other"—produces by itself.

"Additional" absorption by CO_2 = [(1-Tcomb)-(1-TOth)]

The relative additional contribution by CO_2 is:

[(1-Tcomb)-(1-TOth)]/(1-Tcomb)

Table 5 includes the parameters used for the extrapolations of TOth and CO_2.

The "additional" contribution from CO_2 never exceeds 10%.

Table 5: *Additional Absorption*

ppm	TCO2	TOth	(1-TOth)	(1-Tcomb)	[(1-Tcomb) −(1-TOth)]	[...] (1-Tcomb)
5.9	0.8973	0.25	0.75	0.7757	0.0257	3.31%
12	0.8758	0.244	0.756	0.7863	0.0303	3.85%
24	0.8523	0.238	0.762	0.7972	0.0352	4.41%
47.5	0.8278	0.232	0.768	0.808	0.04	4.94%
95	0.8023	0.226	0.774	0.8187	0.0447	5.46%
190	0.7758	0.22	0.78	0.8293	0.0493	5.95%
252.8	0.7645	0.2175	0.7825	0.8337	0.0512	6.14%
380	0.7524	0.214	0.786	0.839	0.053	6.32%
632	0.7283	0.2096	0.7904	0.8474	0.0569	6.72%
1,264	0.6964	0.2036	0.7964	0.8582	0.0618	7.20%
2,528	0.6591	0.1976	0.8024	0.8698	0.0674	7.74%
6,320	0.6023	0.1897	0.8103	0.8858	0.0754	8.52%
12,640	0.5565	0.1837	0.8163	0.8978	0.0815	9.07%

The only way to increase the relative contribution from CO_2 is to increase TOth more than is shown here for CO_2 values below 250 ppm. A critical input would be a specification for K(0) for CO_2 equal, say, to 6 ppm. For the modeling here {TOth & CO_2}, K(0)=284.1K for CO_2=6 ppm appears to be too high. {This is an issue separate to the issues on which this document is focused.}

The dominant role of the background TOth

All of the examples have shown the dominance of the background absorption over the absorption by CO_2. CO_2 alone would never suitably warm our atmosphere, and, when CO_2 is mixed with a stronger absorber, the relative effects of CO_2 are further diminished.

In the previous schematics consider the transition from 252.8 to 632 ppm The slope is ~+0.68C of K(0) per one one-hundredth (1-Tcomb).

If there were no CO_2 and only TOth=0.214, then K(0) would be 284.722. {This is a 9.9C rise from the case of no absorption whatsoever {TOth=1.0; CO_2=0}.} 252.8 ppm of CO_2 increases K(0)=284.722 by 3.1C to 287.821 (**Figure 16**). 632 ppm of CO_2 increases K(0) by another 0.528C. The increase in (1-Tcomb) is 0.0077 and 0.528/0.0077=68.6=0.686C per 0.01 of (1-Tcomb).

Several truths are revealed:
- The background (non-CO_2) absorption is about 3X greater than for CO_2. ((1-0.2140)/(1.0-0.7582) = 3.25.)
- The actual absorption from CO_2 is strongly sublinear for CO_2 concentrations greater than 200 ppm. The most sensitive WN absorption lines have already absorbed over 99% of the BBR flux at those WN (see **Figures 7** and **8**).
- Modifying TOth = 0.214 to the range 0.2187 to 0.2080 as CO_2 is increased from 252.8 to 632 ppm, as shown in **Figure 20**, corresponds to an increase of 0.53C in K(0). The (1-TOth)/(1-TCO_2) ratio decreases from 3.41 to 2.97.
- The 0.0107 decrease (4.9%) in TOth is coincident with a 0.037 (4.8%) decrease in TCO_2.

Progression of atmospheric parameters as CO₂ is increased

When CO$_2$ increases from 252.8 ppm to 632 ppm, a factor of 2.5,[*] very little changes in the atmosphere other than the absorption itself.

Figures 29 and **30** show this progression.

The total absorption of the BBR spectrum, after the BBR has already lost some of its strength to adiabatic loss, increases from 231.16 to 236.83 W/m². The Blkt(0) increase is 2.86 W/m², in exact (required) agreement with the P(0) increase of 2.86 W/m².

K(0) increases by 0.528K (1.0K per 5.42 W/m²).

The ratio, Blkt(0)/CumulativeAbsorption actually decreases as CO$_2$ increases. (See **Table 4**.)

The magnification from the effective (reverse adiabatic) heating effect that adds to this blanket decreases slowly as ppm increases. So there is little effect on the difference in P(0). P(0) increases from

---

[*] LOG2(2.5)=1.3219

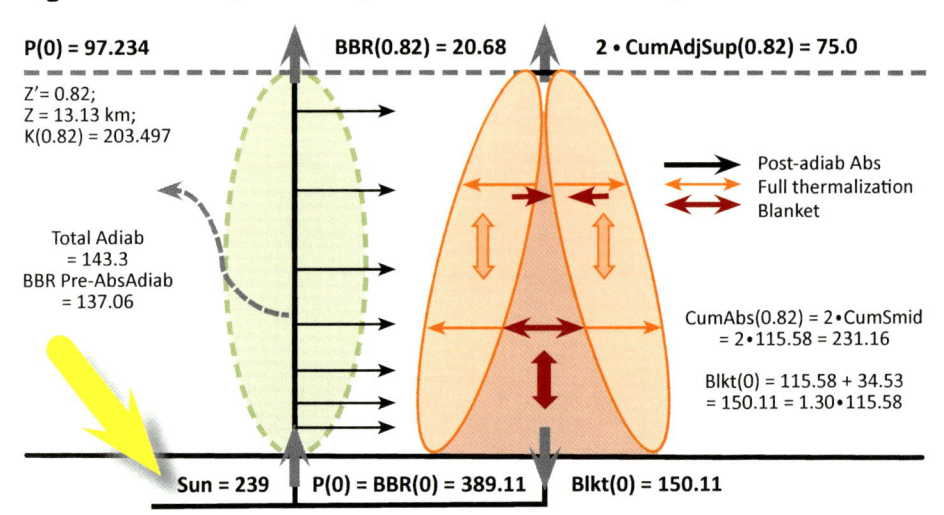

Figure 29: BBR, Blanket, and Adiabatic Flows, Case 1

P(0) = 97.234
Z' = 0.82;
Z = 13.13 km;
K(0.82) = 203.497

Total Adiab = 143.3
BBR Pre-AbsAdiab = 137.06

BBR(0.82) = 20.68

2 • CumAdjSup(0.82) = 75.0

Post-adiab Abs
Full thermalization
Blanket

CumAbs(0.82) = 2•CumSmid = 2•115.58 = 231.16

Blkt(0) = 115.58 + 34.53 = 150.11 = 1.30•115.58

Sun = 239   P(0) = BBR(0) = 389.11   Blkt(0) = 150.11

Energy flows with CO_2=252.8 ppm: TOth=0.214; (1-Tcomb)=0.8351; and K(0)=287.821K.
Note the following consistent conclusions:
 150.11 + (137.06 + 6.24) + (20.68 + 75) = 389.09 ✓
 Adiab (0.82) = 389.11 - 150.11 - (20.68 + 75) = 143.3
 Adiab/Abs = 143.3/231.16 = 0.620

389.11 to 391.97 and K(0) increases from 287.821 to 288.349
 The "escaping" fluxes at the adiabatic cutoff at 13.13 km are almost identical (and their sum is set by the effective pinning of the temperature at Z'=0.82). Some suitable adjustments to the pinning might be recommended, but their effects will be small, since there is only minor absorption taking place at Z' = 0.82 and beyond.
 {See **Appendix E** for plots of many fluxes vs. Z'.}

 $\Delta K(0) = 0.67 \cdot 100 \cdot \Delta(1\text{-Tcomb})$
 $= 0.67 \cdot 100 \cdot (0.8430 - 0.8351)$
 $= 0.529K$
 $0.529K/1.3219 = 0.40K$ per doubling of CO_2

Global Warming Temperatures and Projections

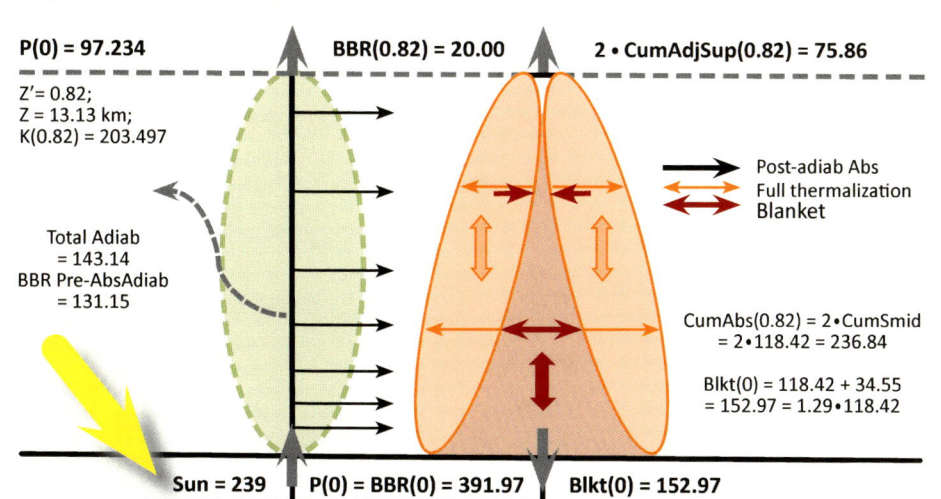

Figure 30: BBR, Blanket, and Adiabatic Flows, Case 2

P(0) = 97.234　　　BBR(0.82) = 20.00　　　2 • CumAdjSup(0.82) = 75.86

Z' = 0.82;
Z = 13.13 km;
K(0.82) = 203.497

→ Post-adiab Abs
⇌ Full thermalization
⇌ Blanket

Total Adiab = 143.14
BBR Pre-AbsAdiab = 131.15

CumAbs(0.82) = 2•CumSmid = 2•118.42 = 236.84

Blkt(0) = 118.42 + 34.55 = 152.97 = 1.29•118.42

Sun = 239　　P(0) = BBR(0) = 391.97　　Blkt(0) = 152.97

Energy flows with CO_2 = 632 ppm: Toth = 0.214; (1-Tcomb) = 0.8430; and K(0) = 288.349

Note the following consistent conclusions:
   152.97+(131.15 +11.99)+(20.00+75.86) = 391.97 ✓
   Adiab = 391.97-152.97-(20.00+75.86) = 143.1
   Adiab/Abs = 143.14/236.84 = 0.604

The gain ratio,

   Blkt(0)/(post-adiab BBR absorption/2)

is basically steady, ranging from 1.30 to 1.28.
   Blkt(0) is about 49% of what the original BBR was before it suffered absorption and adiabatic loss.
   For example:
      150.11/(389.11 - (389.11/ 97.342) • 20.68) = 0.49
and:
      152.97/(391.97 – (391.97/ 97.342) • 20.0) = 0.49
   From 252.8 to 632 ppm it takes an additional 5.67 W/m² of total post-adiabatic absorption to raise the surface temperature by

Atmospheric parameters as CO₂ is increased

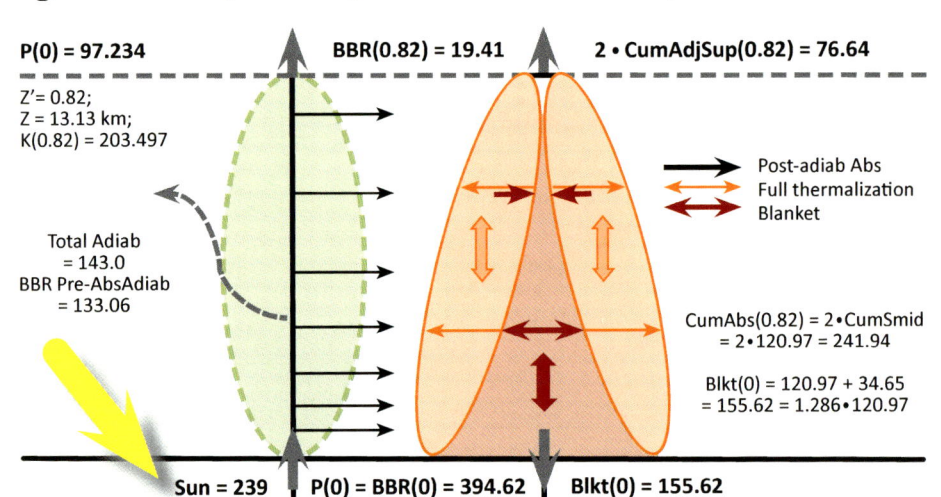

Figure 31: *BBR, Blanket, and Adiabatic Flows, Case 3*

P(0) = 97.234   BBR(0.82) = 19.41   2 • CumAdjSup(0.82) = 76.64

Z' = 0.82;
Z = 13.13 km;
K(0.82) = 203.497

→ Post-adiab Abs
⇌ Full thermalization
⇌ Blanket

Total Adiab
= 143.0
BBR Pre-AbsAdiab
= 133.06

CumAbs(0.82) = 2•CumSmid
= 2•120.97 = 241.94

Blkt(0) = 120.97 + 34.65
= 155.62 = 1.286•120.97

Sun = 239   P(0) = BBR(0) = 394.62   Blkt(0) = 155.62

Energy flows with CO₂ = 1,264 ppm: Toth = 0.214; (1-Tcomb) = 0.8501; and K(0) = 288.834
Note the following consistent conclusions:
  155.62+(133.06+9.94)+(19.41+76.64) = 394.67 ✓
  Adiab = 394.62-155.62-(19.41+76.64) = 143.0
  Adiab/Abs = 143/241.94 = 0.59

0.528K. That corresponds to 10.74 W/m² of total absorption per 1.0K increase of K(0).

Since a 1.0K increase would correspond to an increase of about 5.4 W/m² in P(0), this is a very satisfying "forcing function."

Please note how small the changes are even as CO₂ is increased by 2.5 times. (TOth is dominant.)

Please note the constancy of the adiabatic loss between Z' = 0 and the (pinned) Z' = 0.82. This is what should be expected. The computer simulation also tracks the accumulation of these Adiab values.

The ovals that are used to represent the general thermalization and the specific blanket thermalization after BBR absorption are purely decorative, but they help to define the thermalization effects after absorption. The correct plots (also including the BBR) are shown in **Figures 9** and **10**.

Figure 31 (Case 3, for 1,264 ppm) continues the trend.
Figure 31 has five times the CO_2 of **Figure 29**.
$\Delta K(0) = 1.013K$

1.013K/2.3219 = 0.44K per doubling of CO_2

The situation is extremely stable. A "runaway" is not even conceivable.
The previous calibration parameters are confirmed.

CAVEAT: Figures 29, 30, and **31** are mathematically correct, but, conceptually, they hide the role of water evaporation at the surface (cooling) and its condensation in the atmosphere (heating). More importantly, this local "system" is not self-sufficient since it is ignoring the convection train of heat from the Tropics that is providing drop-offs of Blanket along its passage to the Poles. This train is picking up convection heat from local evaporation as well, but the final AVERAGE Blkt(0) of about 150 W/m^2 is about one-half composed of convection drop-off. The bogeyman of a large Tbkgnd remaining after CO_2 and H_2O absorptions are accounted for is almost entirely comprised of the convection contribution, not some mysterious additional absorbers. In effect, these figures treat this convection *as if it were* a TOth absorption. Section Two, which directly includes water absorbers, and, furthermore, allows for convection drop-offs unaccounted for by the standard spreadsheet calculations, provides the final clarifications of the $CO_2 + H_2O$ story. There may be a better way to present these complicated solutions, but this method has been satisfactory.

Section One summary

A slice model of the atmosphere, with molecular counts rather than altitude as slice "thicknesses," with an averaged temperature for each slice, and with deliberate consideration of absorption and adiabatic loss for WN bands within the Planck spectrum for each slice, has been successfully applied to the question of global warming. The model can address any set of atmospheric absorbers whose density with altitude can be specified, and whose absorption properties as a function of WN (wavenumber) bands can be specified.

A universal calibration plot for determining surface temperature, dependent only on the infrared absorption properties of the atmospheric molecules, has been created.

CO_2 presents no threat to global warming. Its properties are easily modeled. Temperatures, and temperature increases, can be much more easily associated with atmospheric H_2O.

The current absorption parameter (1-T) has a value of about

0.84. It will be increasingly difficult to have this approach 1.0 by introducing more water or more CO_2. More CO_2 alone will have a minor significance; atmospheric water in the form of droplets or clouds is likely to reduce the sun's input because of reflection.

Spreadsheet modeling tracks the upwards-directed BBR (Black Body Radiation) from the surface as the remaining BBR transits each of about 50 slices. Absorptions are calculated across each of about 100 WN bands. Some bands experience no absorption; others have their BBR 99% depleted at low altitudes for CO_2 levels below 250 ppm.

Absorptions create "notches" in the original BBR spectrum, but these notches are filled with isotropic thermalized (non-BBR) photons. Such photons are a component of the Planck thermal distribution, as are all photons from post-absorption emissions.

There is also a tracking (in total power, not as a function of WN) of the quickly thermalized but initially upward-directed half of the post-absorption power flux, and of the initially downward-directed half. The adiabatic losses/gains of these fluxes, within each slice and after leaving each slice, are also tracked. The surface temperature, $K(0)$, is actively influenced by the downward "Blanket" associated with BBR absorptions, but for surface temperatures above 278K, overall absorptions dominate overall adiabatic losses of the BBR flux.

Each $K(0)$ surface temperature requires a specific Blanket flux at $Z=0$. A solution for the equilibrium $K(0)$ is determined only when the model's subtractions, upwards slice by slice, from the required $Blkt(0)$ are found to match the additive results for the downward flux calculated from the cutoff altitude above which temperature and low density essentially eliminate further adiabatic losses. The selected adiabatic cutoff is $Z=13.13$ km ($Z'=0.82$), but results are fairly insensitive to any selection above $Z=11.0$ km.

There is actually no measurable flux downward; the modeling is an accommodation of absorbed BBR to fill the "notches" so as to properly re-create the Planck state of thermal equilibrium.

The pre-absorption loss by the upwards adiabatic loss is perfectly countered by the gain associated with the downwards adiabatic heating.

Global Warming Temperatures and Projections

The combined transmissivity Tcomb for all absorbers is the product of the T's for the individual molecules. The absorption parameter is (1-Tcomb).

The T values are further subdivided as functions of WN, T(WN) and Tcomb(WN). Dissipation from the cooling adiabatic effect is simultaneously considered along with the absorption in each slice.

A calibration plot of equilibrium surface temperature K(0) vs. (1-TOth) for an idealized solo absorber Oth provides the fundamental comparison for all combinations. Oth has the same transmissivity TOth for all WN. Each K(0) requires a specific value for (1-TOth). More specifically, each value of (1-Tcomb) for mixed absorptions has the same K(0) as the matching (1-TOth) for a solo ideal absorber.

In general, Oth means "Other" and (1-TOth) is typically used to represent an unknown (unspecified) background absorption.

TCO_2(WN) has been calculated for all WN for a set of CO_2 ppm. The "effective" overall TCO_2 encompassing all WN is the TOth that gives the same total absorption. For example, the spreadsheet calculations for a CO_2 of 252.8 ppm with its known T(WN), along with a background TOth(WN)=0.2140 for all WN, dictates a K(0) of 287.821K. A solo TOth(WN)=Tcomb(WN)=0.1649 gives the same K(0). Therefore, the effective TCO_2 for 252.8 ppm can always be approximated well by:

Tcomb/TOth=0.1649/0.214=-0.7706

Transmission tables (by Stull et al) for TH_2O(WN) also exist. Overall effective TH_2O calculations could be determined if (a big if) justifiable estimates of specified H_2O ppm as a function of altitude were available. (CO_2 has the advantage of having a well-defined ppm at all altitudes).

Spreadsheet calculations for 252.8 ppm and 632 ppm have been interpolated to obtain a K(0) value of 288K for 380 ppm of CO_2. This baseline point requires, however, an additional background equivalent to a uniform absorber with a TOth equal to 0.2140.

With the assumption that TOth be maintained for higher and lower CO_2 values, it is determined that the $\Delta K(0)$ for each doubling of CO_2 is ~0.42C.

Summary of Section One

The truly important parameter is $\Delta K(0)$ per (1-Tcomb). That relationship is:

$$\Delta K(0) \cong 0.68C/100 \bullet (1\text{-}Tcomb)$$

If T0th is decreased by about 2.8% to 0.2081 (more absorption) for 632 ppm and increased by about 2.2% (less absorption) for 252.8 ppm such that the local linear $\Delta K(0)$ per 100 ppm approximately equals the "official" current estimates of ~0.25C per 100 ppm, then $\Delta K(0)$ for each doubling of CO_2 is ~0.80C.

The background is producing about 3 times the absorption of CO_2, and if it cannot be scientifically linked with CO_2, then it is the background that is the most sensitive parameter. CO_2 itself is not the problem.

The possibility of a "runaway" depends on the value of the "forcing function." The net of all forcing functions should be less than $\Delta P(0)$ per 1.0C for there to be a runaway. It takes an absorption of 10.6 W/m² to obtain a temperature rise of 5.4 W/m² of P(0) associated with a 1.0 increase in K(0). The net of 5.2 says there is absolutely no runaway, CO_2 or no CO_2.

Section One references

Reference publications:

- The Stull et al reference: V. Robert Stull, Phillip Wyatt, and Gilbert N. Glass, "The Infrared Transmittance of Carbon Dioxide," Applied Optics, Vol.3 issue 2, pp 243-254 (1964). The calculations include all quantum mechanical absorption lines. Tables are provided for a specified set of CO_2 molecules per square centimeter (convertible to ppm). The table values are the integrated transmission coefficients for bands of wavenumber (WN) centered on WNmid and with a WN width of $\Delta WN = 50$ cm^{-1}. (The author's document modifies each table cell to each specific atmospheric slice layer and its appropriate temperature.)
- http://farside.ph.utexas.edu/teaching/sm1/lectures/node56.html. These are class notes on the adiabatic effect prepared by Prof. Richard Fitzpatrick of University of Texas-Austin.
- www.barrettbellamyclimate.com. This web site discusses Schwarzschild's equation in an appropriate differential mode. (The approach for the calculations in the author's document is

a common sense construction that aligns with the more complete Schwarzschild development, and employs WN (wavenumber) rather than frequency, transmission parameters rather than absorption parameters, direct corrections for the adiabatic effect, and summations rather than integrations.)

- Trenberth et al, http://www.cgd.ucar.edu/staff/trenbert/trenberth.papers/BAMSmarTrenberth.pdf. (Trenberth et al have a series of influential documents. The baseline input for the "average" state in the author's model is chosen to be identical to the average state for Trenberth et al. From this baseline, the author's atmospheric slice approach of duplicates all of the final outputs for Trenberth et al. **In addition, the author's approach allows projection calculations to be made for changes in the input parameters.)**

Section Two:

H₂O and its phases

H_2O water vapor influence on global warming vis a vis CO_2

The previous sections have discussed and resolved the major issues of CO_2 as an absorber and as an influence on K(0). A universal calibration plot has been established, one that can incorporate all absorbers with WN-dependent transmissivity and can represent unknown "background" absorbers as an absorber Other with uniform transmission at all WN. For any mix of absorbers, a composite absorption parameter (1-Tcombination) = (1-Tcomb) supplies the equilibrium value for K(0). Water has its own parallel contribution with all other absorbers.

There are assumptions. The modeling by the international team, designated as Trenberth et al, provides the basic baseline average conditions. Their modeling applies a spatial and temporal worldwide averaging of the parameters of insolation, cloud absorption and reflection, earth reflection, and atmosphere and earth absorptions. In the author's modeling, adiabatic effects are specifically included. Since the author's modeling covers a relatively wide range of K(0), the "lapse rate," the rate of falloff of K(Z) as altitude Z increases, is assumed to have a "pinning point" at Z = 13.13 km. K(Z = 13.13 km) = 203.5K. P(Z = 13.13 km) = 97.2 W/m² as K(0) changes.

Observational techniques (the necessary confirmation of predictive models) tend to use TOA (Top Of Atmosphere) data gathering. Since the observations automatically include K(0), adiabatic losses can be inferred only for the observed cases. TOA altitudes are always greater than the adiabatic cutoff altitude, and tend to be at or above 20 km. The author's modeling is specifically concerned with absorption contributions to the Blanket as a function of altitude, and so the cutoff altitude is a critical component. The author's modeling involves only integrations with altitude and so lateral convection issues require separate consideration. **In the previous calculations convection contributions to the Blanket are buried within the TOther and Tbkgnd terminology.** The convection issues are intimately associated with water condensation and are addressed in this section.

The specific transmissivity of water vapor was not included in the previous calculations because a) those discussions were centered on the role of CO_2 and b) no baseline average for water distribution as a function of altitude has ever been recommended. Nevertheless, WN-dependent transmission parameters for water vapor exist and can be incorporated into the previously described Tcomb modeling if an appropriately averaged distribution with altitude can be supplied. "Appropriately averaged" is a very ambitious goal since water levels can change by an order of magnitude within hours.

Water distributions with altitude can, however, be introduced into the spreadsheet calculations. A wide range of water distributions in the average (temperate) zone have been introduced into the spreadsheet calculations, and there is, indeed, an increase in the blanket associated with both absorbers (CO_2 and H_2O). Those results fit perfectly onto the calibration plot, but there is still an unknown TOth background that is larger than should be expected.

The purpose of this Section is to demonstrate there is an additional contribution to the blanket that is NOT associated

with local absorptions, but that is, nevertheless, associated with water. This association with water has no clearly definable correlation with CO_2. *The official IPCC models have continued to assume a significantly incorrect forcing function* by which the $\Delta K(0)$ associated with the additional CO_2 creates more water, more water absorption, and a blanket increase of 2 to 2.5x with even small increases in CO_2.

The effects of water — its evapotranspiration from water and plants, and its reconstitution as water (and ice) at higher altitudes as clouds — and the effects of the clouds' albedo and absorption are topics of much greater depth than the issues associated with CO_2. The previous sections have already shown that "Other" (water, presumably) is a much stronger contributor to $K(0)$ than is CO_2.

This section eventually discounts the assertions that any temperature increases caused by CO_2 are significantly enhanced because of a co-generation of water vapor, which in turn produces more absorption, a higher temperature, etc., etc. Water is, in fact, a temperature moderator, not only because of its high heat capacity in the ocean (and the ocean's ability to distribute heat) but also because of the *fundamental thermodynamics of evaporation from the earth and the eventual condensation of water vapor in the atmosphere*. For whatever indirect effect there is, it cannot be simply approximated by a multiplicative factor as if it were direct.

The adiabatic lapse rate

The lapse rate is the rate at which K(Z) falls off as the altitude Z increases. The tag 'adiabatic' means that no outside heat is involved during any cooling (rising of air) or heating (falling of air). Moist air, even a few percent, has a higher heat capacity and its lapse rate — its gain or loss of heat with falling or rising — is lower than for dry air. **A composite gram of water vapor releases 1.0 Calorie of heat for each 1.0C temperature drop**; a composite gram of dry air releases about 0.24 Calories for each 1.0C; a typical gram of "moist air" releases about 0.243 Calories for each 1.0C.

Completely dry air has a lapse rate of minus 10C per kilometer, but the more typical value is minus 6.5C/km for moist air.

But the lapse rate isn't fundamentally about moisture. After all, the lapse rate is greater for dry air. An isothermal (or isopressure) atmosphere is not even possible in a gravitational field.[*] A water

---

[*]Doug Cotton and Clive Best, clivebest.com/blog/?p=4517.

molecule is about two-thirds the weight of a dry air molecule, and is less constrained by gravity. The thermodynamic search for equilibrium requires particles at higher altitudes to have higher potential energy and lower kinetic energy than those at lower altitudes. The lower kinetic energy means a lower temperature. And the lower density and lower temperature at higher altitudes also guarantee lower pressures. Given the molecular contents of the earth's atmosphere, and an absence of clouds, the atmosphere's equilibrium density with altitude and its temperature with altitude can be accurately calculated. Direct measurements provide confirmation of such calculations.

Water vapor content, clouds and rainfalls are the most difficult variables, and attempts to provide averages always seem to require averages of measurement data.

The evapotranspiration of water is not tightly linked to higher surface temperatures *caused by CO_2 absorptions* within the atmosphere. The atmosphere accommodates the full range of relative humidities (RH). Higher temperatures may have identical absolute humidities (AH) to the AH at other locales with lower K(0), but may have reduced RH. **And the act of evapotranspiration actually reduces surface temperature.** When water is evaporated from the earth **it takes heat** from the earth in order that the water be vaporized. This is the same effect as the cooling of the skin when water evaporates. **The earth is cooled by the evaporation.** As the water vapor molecules rise, the moist air loses 0.24 to 0.25 Calories per gram for each 1.0C (or K). At higher altitudes (of several kilometers) in the atmosphere the dew point is reached and water droplets are formed. **The heat of condensation that is released is 540 Calories per gram. Each gram of water (about 3.3E22 molecules) is heating the atmosphere by this amount.**

This release of heat well up into the atmosphere disrupts the lapse rate. It acts as an independent, but spasmodic, heat source. This heat is distributed laterally and vertically by convection. Cooled water if/when it falls to earth further cools the earth. There is no runaway effect. This recycling is a self-tempering act of nature. **It would appear to be more logical to propose a thesis that evapotranspiration acts to** *reduce* **the surface heating**

caused by the solo effect of BBR absorption in the atmosphere. Of course, any maintained residence of water in the atmosphere between rainfalls will increase K(0) in the same manner as for BBR absorption by fixed CO_2.

The shift of heat from the earth via water vapor to an upper level of the atmosphere via vapor condensation is an ideal transfer. Because of convection much of this heat never returns to the earth area below where the vaporization took place. CO_2 or no CO_2, that portion of earth is cooled.

The convection activity associated with the condensation in the atmosphere contributes significantly to weather variability. Any lateral and vertical impulses in atmospheric pressure further contribute to the weather/climate turbulence. But none of this has any relation to CO_2 itself.

Specific comparisons of atmospheric water and CO_2

The onus of CO_2 is that, once welcomed into the atmosphere, it has a long lifetime. Each CO_2 molecule can continue to absorb BBR fluxes and thereby create blanket fluxes. Water molecules in the atmosphere may have a lifetime of a week or more because any significant buildups result in rainfall, but the half lifetime, or residence time, of CO_2 in the atmosphere is 20 to 40 years. The residence time of CO_2 is not a reference to individual molecules but only with a steady state value; considerable exchange continuously takes place with the land and especially with oceans and lakes. The daily release by the oceans far, far exceeds the daily increase of CO_2 from human activity, but the oceans each day have approximately the same input from the atmosphere. On the other hand, the residence time of water in the atmosphere is just over one week.

The CO_2 solubility of the oceans is directly observable during the periods of El Niño and La Niña. El Niño produces warming and the CO_2 levels increase more than any current trend from human activity. This is caused by the lower solubility of warmer oceans for CO_2 and their net release of CO_2. During La Niña, the oceans are cooler and willingly accept more CO_2 from the atmosphere, even to the extent of atmospheric CO_2 levels flattening or decreasing. These CO_2 relations are strongly observed only in the northern hemisphere.

Rainfall also provides the blessing of purging the atmosphere from particulates caused by either nature or man. CO_2 is not harmful at all to human health but particulates of all kinds are especially harmful. This does not justify particulates from carbon burning, but emphasizes the health value realized when particulates are removed, even if accompanied by more CO_2.

The present weight of CO_2 in the atmosphere is about 3,400 gigatons. The present estimate of water in the atmosphere is about 3,100 cubic miles, or 14,000 gigatons. The water in the atmosphere is about 0.001% of the earth's total water. If all of the atmosphere's water were spread evenly over the earth (including over the oceans), its thickness would be about 3 centimeters. Locally, however, it can be greater than 30 cm, as we know from some hurricanes! It is water vapor variability, its lateral and vertical convection and movement by wind, temperature shifts caused by wind and ocean flow, and earth-rotational triggered impulses in atmospheric pressure that produce those climate-related catastrophes not associated with tsunamis, earthquakes, and volcanoes. CO_2 is not involved.

The constantly asked question is: Do CO_2 increases precede temperature rises or do temperature rises precede CO_2 increases? **The onset of Ice Ages demonstrates decreases in atmospheric CO_2, which is consistent with cooling oceans steadily purging the atmosphere of CO_2.** The gradual exit from an Ice Age is accompanied by gradual increases in CO_2. This, undoubtedly, is the release of CO_2 from the oceans. But the initial CO_2 was almost certainly from carbon burning, by nature and, eventually, from nature's biological life.

Cloud influences are considerable and are imperfectly modeled. They can reflect sunlight and reduce warming. The cloud's water will absorb BBR, especially at night, and will raise temperature. The formation of water droplets releases heat into the lower

Global Warming Temperatures and Projections

atmosphere and the heat is carried elsewhere by convection. This has a partial cooling effect on the surface. The fractional amount of total water that is in the clouds is, however, quite small.

BBR absorptions by water molecules or CO_2 molecules are, however, fundamentally the same. The spreadsheet calculations for either water vapor or CO_2 are identical; each has its own WN sensitivity. The final $K(0)$ depends on the combination of CO_2, water, and all other absorbers. **For equivalent ppm, the effective transmissivity of CO_2, *eff*TCO_2, is lower than for H_2O, but the higher ppm, evaporation, condensation, convection, and non-local blanket contributions from the convection result in "water" being responsible for about 75% of the average blanket.**

The fundamental difference is that CO_2 is essentially uniform, globally and with altitude. CO_2 concentrations have been increasing slowly with time. It is readily amenable to computer calculations such as those presented in the earlier sections. Water vapor concentrations differ in all locales in the world, change dramatically from day to day, and can also vary with altitude. **There is absolutely no immediate linkage with CO_2, and no demonstrated long-term linkage either**, whether by scientific argument or data observation. An obvious question is whether or not average water concentrations have increased with time.

Average water vapor content in the atmosphere

Common measurements of localized water content are: water vapor pressure (hPa, hecto-Pascals); Absolute Humidity (grams of water per cubic meter); wet specific humidity (grams of water per kg of air); Relative Humidity (% of the maximum possible absolute humidity); and ppm (parts per million). The vertical integration of all the water in the atmosphere is known as the precipitable amount (cm) or TPW (Total Precipitable Water). An alternate expression for TPW is (integrated) kg/m^2 of water. A TPW of 2 cm equals 20 kg/m^2.

There are two versions for ppm, ppm_v (ppm per volume equivalent of air, or molar fraction), and ppm_w (ppm per weight equivalent of air, or mass fraction). $ppm_w = (18/29) \cdot ppm_v = 0.621 \cdot ppm_v$. Whereas ppm ($ppm_v$) for CO_2 are close to 390, ppm for water easily range from 1,000 to 20,000 (or higher). **A precipitable water count of 3 cm**, *if the AH were constant with altitude*, **corresponds to** ppm_v = **5,120 and** ppm_w = **3,180.** 2.5 to 3 cm is the average water content in the atmosphere, but, locally, its integration with altitude can exceed 30 cm. Water ppm_v can vary from 100 to 100,000.

Some brief derivations follow (see also **Quick references (5**

tables) following the appendicies):
- There are 1E44 molecules in the atmosphere. The area of the earth is 510E12 m². **The areal density of the earth molecules at Z = 0 is 1.96E29/m².** At 239 W/m² the sun's input power to the earth is 122 billion gigawatts.
- The density of dry air is 1.24E3 grams/m³. By employing the Avogadro number, 6.022E23 molecules per mole, and the average molecular weight of 29 for air molecules, **the number of molecules of dry air at STP (standard Temperature and Pressure) per m³ is calculated to be 1.24E3 • (6.022E23/29) = 2.575E25 molecules/m³.**
- The next section references actual NASA data; that data indicates no precipitable water beyond 9.166 km (Z' = 0.697). Therefore, **only the count below Z = 9.166 km (Z' = 0.697) is of importance in determining precipitable H_2O. That count is 0.697 • 1.96E29 = 1.366E29 molecules/m².**
- Let us use 1,000 ppm$_v$ of water vapor for all future calibration determinations. 1,000 ppm$_v$ of water vapor = (1,000/1E6) • 2.575E19/cm³ = 2.575E16 water molecules/cm³.
- ppm$_w$ = (18/29) • ppm$_v$ = 621. Even within the limited range of Z' = 0 to Z' = 0.697 1,000 ppm$_v$ is still 2.575E16 water molecules/cm³.
- Therefore, **for ppm$_v$ = 1,000, ppm$_w$ = 621. The ratio is 1.610.**
- **The particular estimate of cm of precipitable water for 1,000 ppm$_v$ distributed uniformly to Z' = 0.697 = (1,000/1E6) • 366 • (18/6.022E23) = 0.408 cm.**
- The average situation of 3.0 cm of precipitable water (to a height of 9.166 km) has a water ppm$_v$ of approximately 7,350. So water in the atmosphere exceeds CO_2 by much more than a **factor of 10.**

The Stull et al reference ("The Infrared Transmittance of Water Vapor," by P. Wyatt, V. Stull & G. Plass. Applied Optics, vol. 3, No. 2, Feb 1964) has WN-dependent absorption tables for 2, 5, 10, ... centimeters of precipitable water. K(0) calculations, with and without CO_2, are discussed later.

Published distributions of water vapor

The distribution of water vapor in the atmosphere has been studied by climate scientists for years. The following publications are of particular note.

- Clive Best's plots of precipitable water vapor: clivebest.com/blog/?p=4517
- Leif Svalgaard's Weather and Climate Analyses Using Improved Global Water Vapor Observations: (http://www.leif.org/EOS/2012GL052094-pip.pdf)
- Friends of Science Climate Change Science Essay (https://friendsofscience.org/index.php?id=681) with a section on water vapor feedback (www.friendsofscience.org/index.php?id=710).

Figure 32, reproduced from Friends of Science with permission on the following page, gives the average precipitable water vapor by atmospheric layer. The average water vapor is shown in millimeters as measured by NASA from the years 1988 through 2001 for

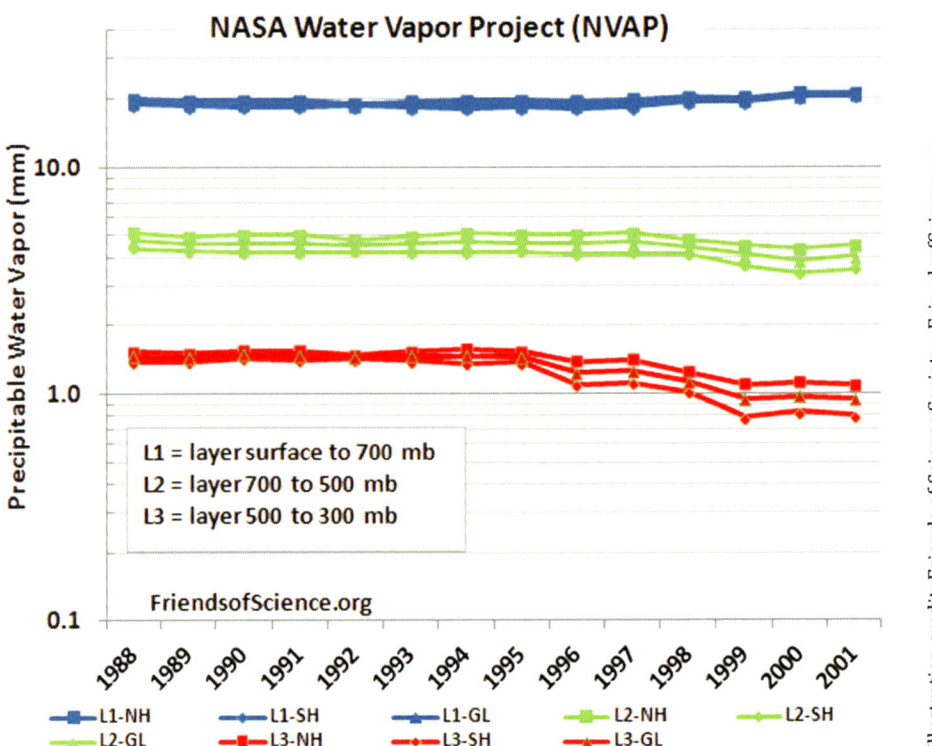

Figure 32: *Precipitable Water Vapor by Layer*

the Northern Hemisphere, the Southern Hemisphere and the entire globe. There are three "Levels," corresponding overall to the pressure range from 1,013 millibar ($Z = Z' = 0$) to 300 mb ($Z = 9.166$ km; $Z' = 0.697$). The measurements are taken for each of three layers, earth (1,013 mb to 700 mb (about 3.0 km), 700 mb to 500 mb (~5.6 km) and 500 mb to 300 mb (about 9.2 km).

Level 1 from 1,013 millibar ($Z = Z' = 0.0$) to 700 mb ($Z = 3.01$km; $Z' = 0.35$) contains about 50.2% of the total number of molecules between 1,013 mb and 300 mb ($Z = 9.166$km; $Z' = 0.697$). It contains 42.7% of the molecules between $Z = 0$ and $Z = 13.13$ km ($Z' = 0.82$; 162.8 mb).

Level 2 from 700 to 500 mb ($Z = 5.576$ km; $Z' = 0.539$) contains about 24.1% of the molecules between 1,013 mb and 300 mb. It contains 20.5% of the molecules between $Z = 0$ and $Z = 13.13$ km ($Z' = 0.82$).

Level 3 from 500 to 300 mb ($Z = 9.166$ km; $Z' = 0.697$) contains about 25.7% of the molecules between 1,013 mb and 300 mb. It contains 21.8% of the molecules between $Z = 0$ and $Z = 13.13$ km ($Z' = 0.82$).

The actual results from NASA data give a total average water content of approximately 2.6 cm. The {Lev1, Lev2, Lev3} results of {2.0, 0.45, 0.15} can be stated in percentages as {77%, 17%, 6%} even though the total molecular percentages for these three ranges are {50.2%, 24.1%, 25.7%}. There is a fourth level, from 9.166 km to 13.13 km, which effectively contributes 0 cm (and 0%).

The discussion in the previous section was assuming that the absolute humidity of water vapor or liquid water was constant with altitude (i.e., for Z' altitude, for which any given $\Delta Z'$ always represents the same total count of molecules). This is, of course, ultimately unrealistic, since the rising water vapor, with its rising RH, does not get past the actual water or ice stages, except in so far as water and ice have their own partial vapor pressures associated with the altitude's temperature and overall pressure. For example, with a 50% surface RH at a surface of temperature of 288.0K the dew point is reached below 3 km. The Level 1 range is from 0 to 3.01 km ($\Delta Z' = 0.350$) and so Levels 2 ($\Delta Z' = 0.189$) and 3 ($\Delta Z' = 0.158$) will contain water/ice concentrations. A fourth level from $Z = 9.166$ to 13.14 km ($\Delta Z' = 0.123$) can be ignored.

This is only one of the reasons why the CO_2-related discussions had so much difficulty in including a "standard" water distribution within the modeling.

These three actual Ln values can be folded into the previous section on "average water calculations in the atmosphere." For 1,000 ppm, the precipitable water calculation for $Z' = 0$ to $Z' = 0.35$ (Level 1) would have been

(1,000/1E6) • (0.35 • 1.96E29) • (18/6.022E23) = 0.205 cm

The NASA data value of 2.0 cm for this range converts to about **9,800 ppm$_V$**.

Figure 33: *Global Relative Humidity at selected altitudes over a period of 60 years*

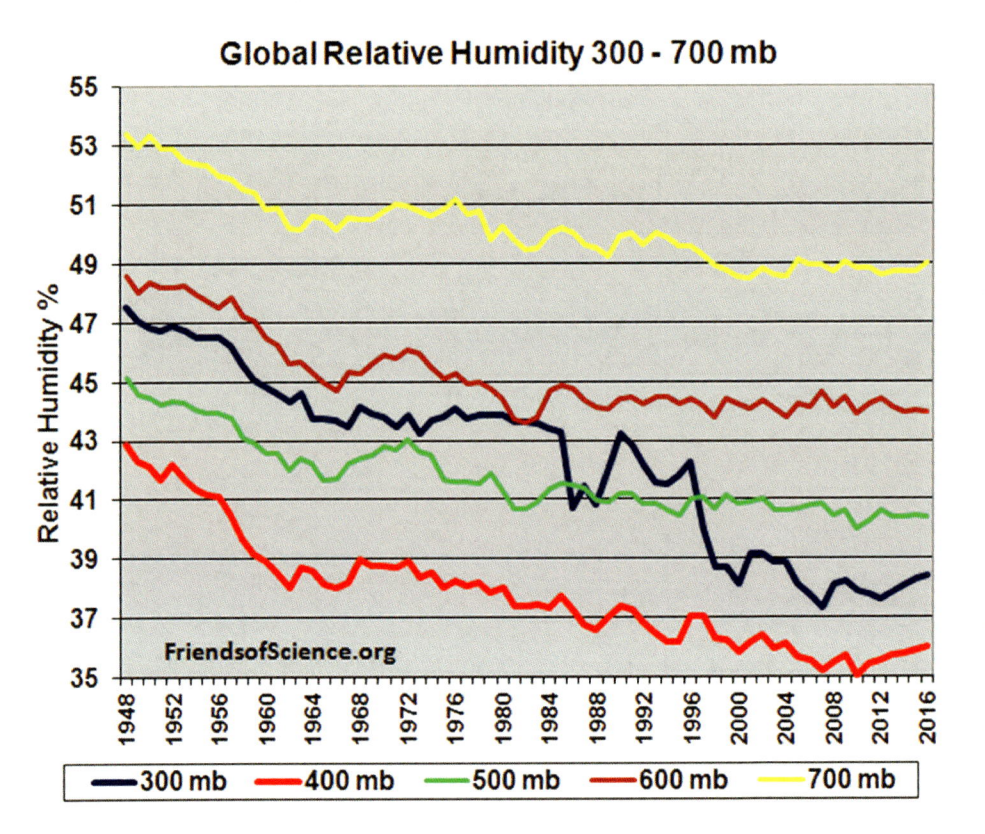

A working assumption therefore, is that the L1 2.0 cm observation by NASA corresponds to a ppm$_v$ at the surface of 9,800. An additional, and more empirical, assumption would be to replace the composite **(0.347 • 1.96E29)** for the combined Levels 2 and 3 with **(0.3) • (0.347 • 1.96E29)** and to set **(0.123 • 1.96E29)** for level "4" equal to 0. That is, reduce the sum of calculations between 700 mb and 300 mb by a factor of 3.3. This concatenation gives the final count of 2.6 cm. This correction factor of 0.3 can be improved by means of a calibration table that incorporates K(Z = 0) and RH(Z = 0) as its inputs. The dew point altitude is the critical output in determining

Figure 34: *Specific Humidity (grams of water per kg of air) at an altitude of about 8 km over 60 years*

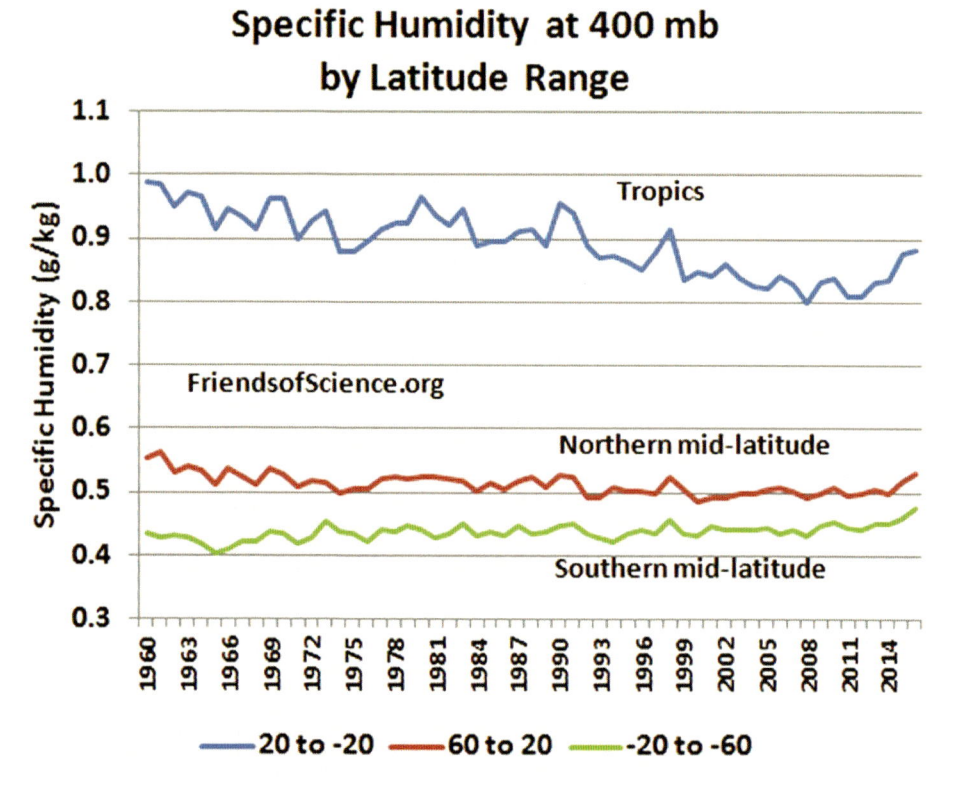

the local TPW.

A recent paper *Challenging climate sensitivity: 'Observational Quantification of Water Vapor Radiative Forcing,'* by W. Eschenbach and A. Watts at the winter meeting of the American Geophysical Union, employs satellite data that demonstrate an increase of 1.5 kg/m² (a TPW increase of 1.5 mm) over the past 28 years, with the overall mean being 28.7 kg/m². This correlates with an increase in absorption and blanket. No correlations with CO_2 were presented.

A typical assumption by anyone who claims the atmospheric temperature is increasing and that this must necessarily be accompanied by an increased AH at Z = 0 because of increased water content is that the Relative Humidity for higher Z would either stay the same or increase. The IPCC models assume RH stays constant with altitude. The data confound this seemingly logical statement. **Figure 33** shows the RH at five levels in the atmosphere for a period of 65 years. RH is lower for each higher altitude except for the highest altitude (300 mb). Additionally, **RH at all altitudes is relentlessly decreasing with time.**

The reduction, especially at higher altitudes, allows more heat to escape. The globally averaged curve for 300 mb (the highest of the altitudes — about 9.16 km) is dominated by the tropical zone "crown" of water that pushes to higher altitudes. **Near the surface, however, at 1,000 mb, the average global RH has been constant at about 78%.**

Figure 34 shows 65 year trends of specific humidity at an altitude of 400 mb (about 8 km) for three regions of latitude. **Specific humidity (grams of water per kg of total air) is a variant on absolute humidity (grams of water per cubic meter).** The specific humidity for the tropics is approximately twice that for the mid-latitudes and is steadily decreasing. The specific humidity for the North and South mid-latitudes is nearly constant. Separate plots (not presented here) show that the decrease in water vapor over the past 65 years is particularly dominant in the 300 to 500 mb range, and that this is true for all latitudes. If AH (or specific heat) is constant and RH is decreasing, then temperature must be increasing. If AH is decreasing and RH is decreasing, then temperature must be decreasing.

Figure 35 gives a plot of the global average of specific humidity at 400 mb, but the X-scale, which had been given in years, is now being represented by the official average ppm CO_2 value for each year. A second plot shows the prediction of the current official model for specific humidity vs. CO_2. The divergence is significant. Despite the flat model predictions for a range of CO_2 increasing by 80 ppm both the global and the tropics data show significant reductions in specific humidity. The global reduction is about 10%; the tropics reduction is about 14%.

The basic problem seems to be the official model's assumption

Figure 35: Specific Humidity (grams of water per kg of air) at an altitude of about 8 km vs. CO_2 concentration over a period of 60 years

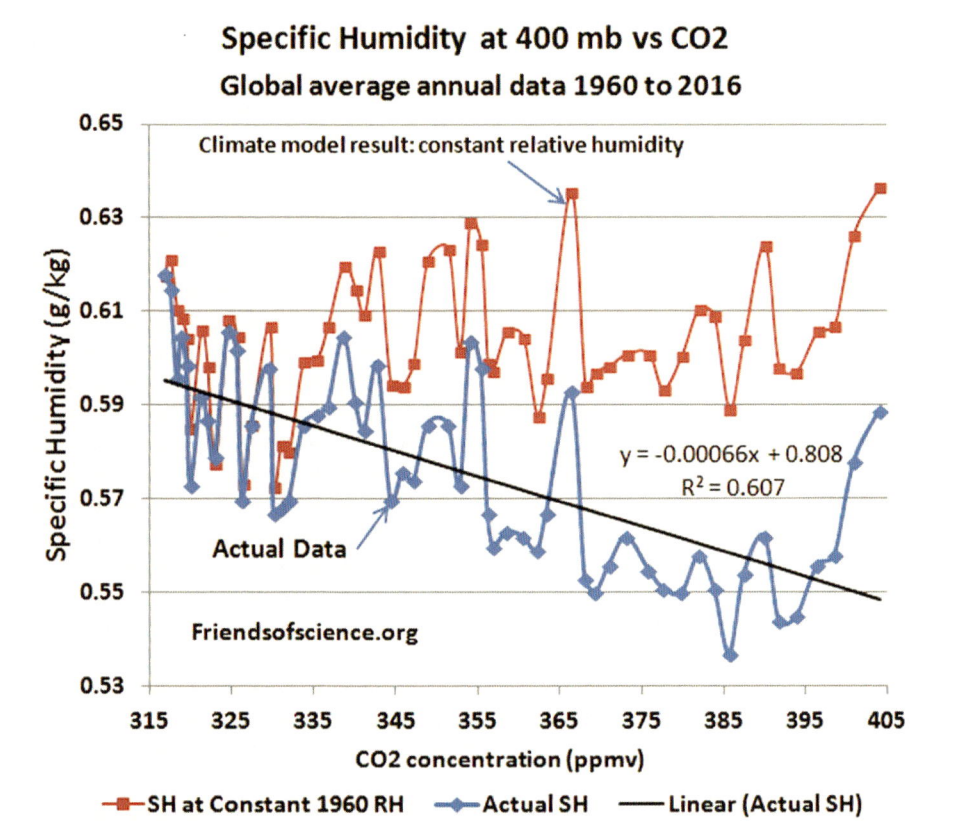

Specific Humidity at 400 mb vs CO2
Global average annual data 1960 to 2016

Climate model result: constant relative humidity

Actual Data

$y = -0.00066x + 0.808$
$R^2 = 0.607$

Friendsofscience.org

— SH at Constant 1960 RH   — Actual SH   — Linear (Actual SH)

Illustration credit: Friends of Science Society - FriendsofScience.org

of constant RH. **Figure 33** has already demonstrated the reduction of RH at all altitudes.

Figure 36 gives the precipitable water for a single year (1991) as a function of latitude. (**Figure 32** gives the averaged precipitable water for a sequence of 14 years.). The falloffs on either side are predictable. The "crown" for the tropics is truly a crown since the water extends much higher into the atmosphere. The southern

Figure 36: *Precipitable Water Vapor in millimeters vs. Latitude for a specific year*

Precipitable Water Vapor by Layer in 1991

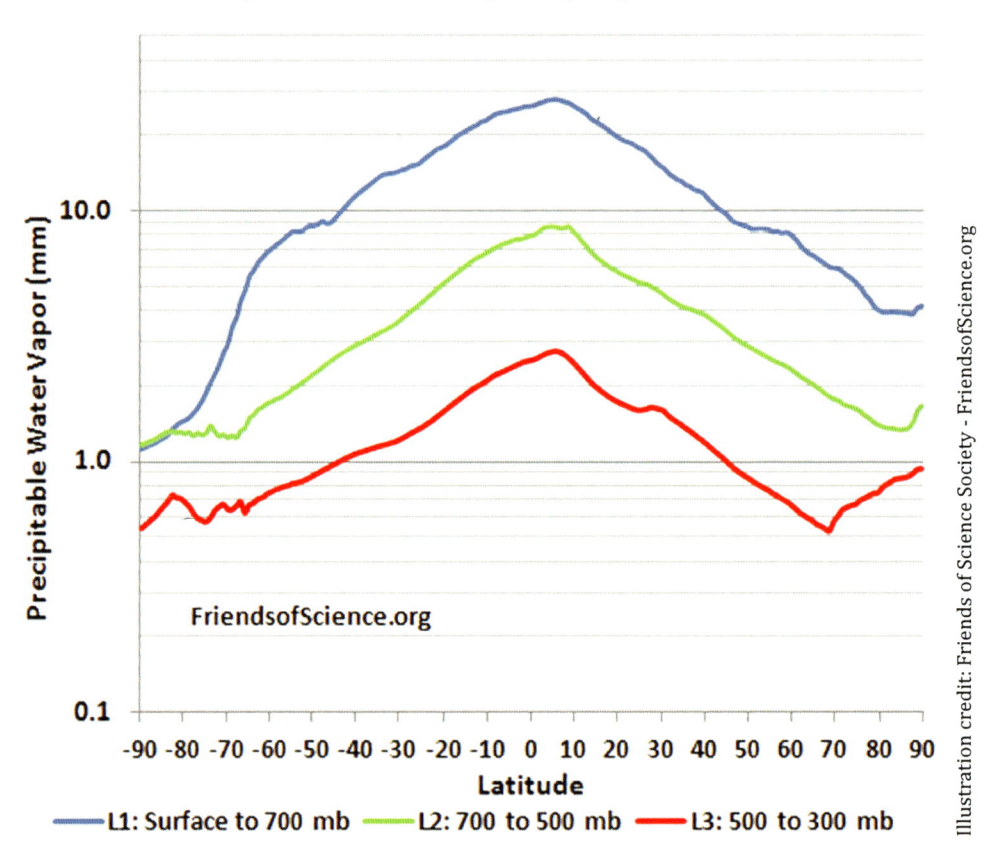

FriendsofScience.org

— L1: Surface to 700 mb   — L2: 700 to 500 mb   — L3: 500 to 300 mb

Illustration credit: Friends of Science Society - FriendsofScience.org

hemisphere has more ocean water and is buttressed by the frozen Antarctic continent with its miles high of ice and snow. The pole of the northern hemisphere has no continent and consists of a relatively thin snow and iceberg that is undercut by a relatively warm ocean stream that varies from month to month and year to year.

Much of the warmth in the higher (non-tropic) latitudes is the result, in fact, of lateral convection and physical transport from the

Global Warming Temperatures and Projections   **137**

Figure 37: *Averaged Temperature vs. Latitude at ten year intervals over a period of 60 years.*

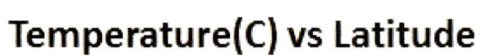

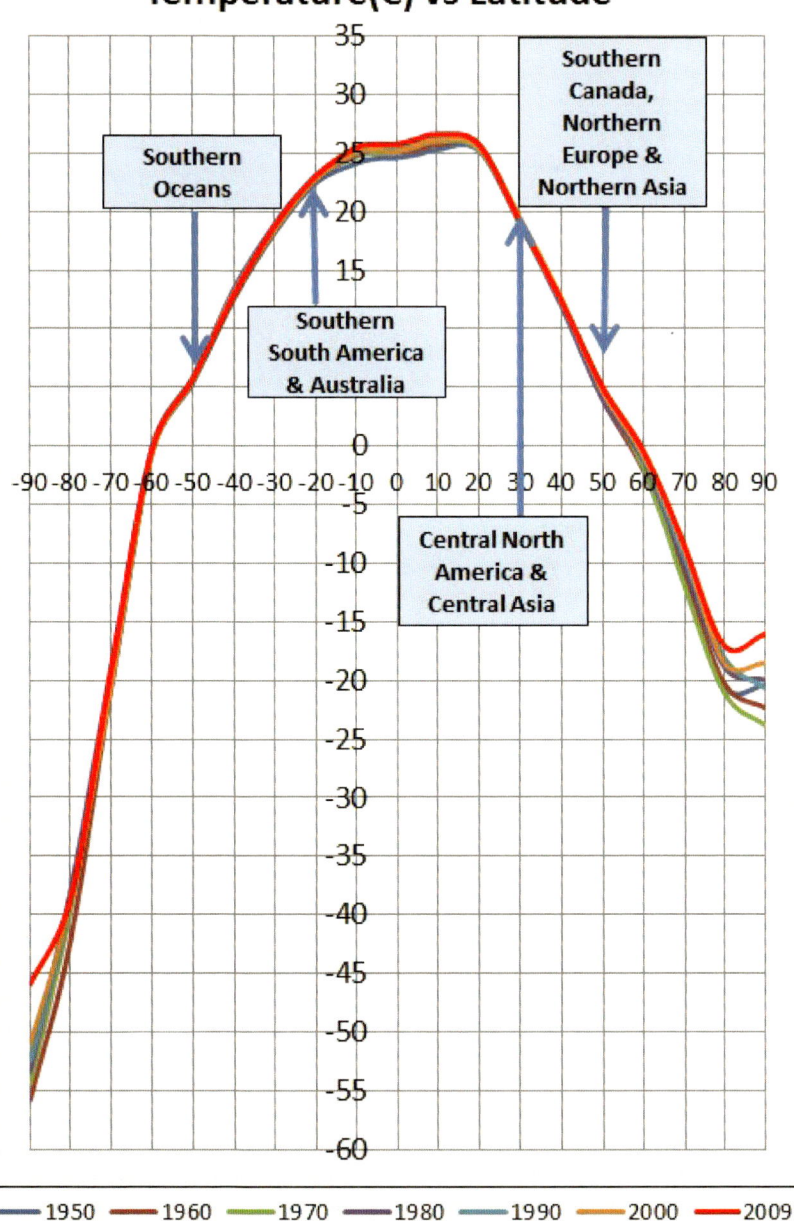

tropic regions. There will be some specific comments about that in the next section

Another highly recommended reference is: www.roperld.com/science/PrecipLatitude_Longitude.htm

It provides overlays of 60 years of plots (seven at ten-year intervals) of: Precipitation rate vs. Latitude (**Figure 44**); Precipitable water (kg/m^2) vs. Latitude; and Temperature vs. Latitude (**Figure 37**).

The overlays are nearly perfect, and the differences between the northern and summer hemispheres are clearly delineated.* The Southern hemisphere has more southern oceans and the huge tonnage of ice and snow on the Antarctic continent. The North pole has warm ocean streams under its floating ice and snow. Temperature increases appear to be near zero for central North America, and less then 0.5C for the worst segments of the tropics. Neither CO$_2$ nor water appears to be a problem. It is difficult to conceive of runaway possibilities after looking at these three figures from roperld.

This Roper paper does not indicate whether its temperatures are accurate averages across the entire area of the globe including the ocean areas. The fanout at the North Pole is due to variations in ocean currents under and around the ice, and the lower temperatures at the South Pole are due to the massive continent of ice and the expansive distances to the cold surrounding ocean water. However, overall near-surface ocean measurements since 1980 indicate a temperature increase of about 0.5C, and they may be the best representative (if carried out completely and accurately) for Δglobal temperature. A ΔK(0) of 0.5C corresponds to a ΔP(0) of 2.7W/m^2 and to a ΔBlkt(0) of 2.7 W/m^2. The ΔBlkt(0) of 2.7 W/m^2 corresponds either to a ΔMolecularAbsorption of 5.4 W/m^2 **and/or** to a ΔConvection (downward) of 2.7 W/m^2.

---

* The reasons for the offsetting peaks and averages in the southern hemisphere are well-explained by Professor Richard Seager in his on-line document http://ocp.ldeo.columbia.edu/res/div/ocp/pub/seager/Kang_Seager_subm.pdf

Heat convection from the Tropics

As a continuation of the discussion on precipitable water and a recycle time of nine days, consider the average power released by the latent heat of condensation in converting vapor to water. **Consider a TPW of 3 cm of precipitable water being created and being dispersed every nine days** (3.3 mm/day); one can scale for any other thickness divided by time span. The power for the heat released into the atmosphere equals the latent energy per gram times the grams per second.

Power of condensation in W/m^2 =
(2,257 Joules/gram)(3E4 g/m^2)/((3,600 • 24 • 9) s)
= 87.1 W/m^2

This slab of heat spreads by convection laterally and vertically. Much of it will arrive (somewhere) at the surface as part of the blanket but the component directed upward has a better chance of escape/adiabatic loss than the same amount of surface BBR emission would have of being absorbed in the lower kilometers, and of subsequently contributing to the blanket.

Figures 38, 39, and **40** are useful schematics of the effects of

Heat convection from the Tropics

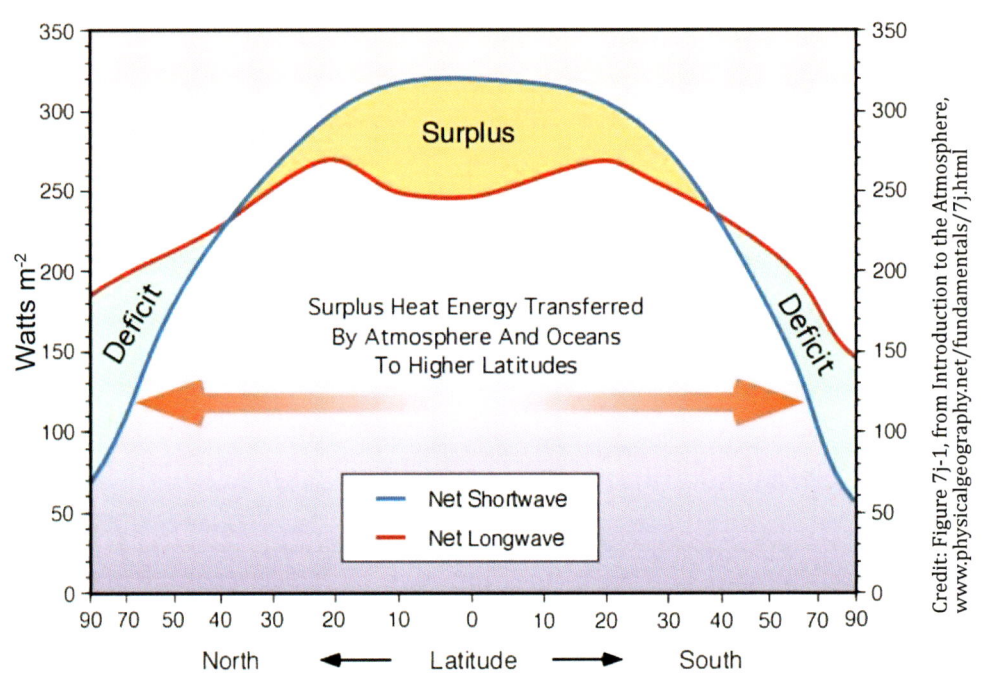

Figure 38: *Balance between average net shortwave and longwave radiation from 90 North to 90 South*

convection. **Figure 38** shows the average situation across the entire 180 degrees of latitude. (The N Pole is to the left for this particular plot.) The tropics extend from S 23.5 degrees to N 23.5 degrees. The sun is particularly effective in this "wobble" region of the earth's spinning. **The "crown" of heat relative to the actual earth emission in the Tropics is significant**; the difference across the Tropical region is about 75 W/m². **The associated power of 75 times the tropical area convects upward and laterally towards the two poles.**

Both poles receive a significant amount of additional heat as does the temperate region. The symbolic convection is shown, in a simplistic manner, in **Figure 39**. {This simplistic picture will eventually be enhanced within a discussion of the actual circulation loops (tubes) within each hemisphere.} There is more water held, on average, in the tropics and it extends higher in altitude. **The down arrows in the temperate region represent additional blanket that has not been comprehended within the absorption measurements of the spreadsheet.** *It has been "comprehended" only in the sense that it reveals itself as an unspecified "background."*

K(0) in the Tropics is 14C lower than it would be if all the heat of condensation had been returned to the earth. The temperatures at the poles are 15C to 25C higher than they would be otherwise. This is just another marvelous example of the smoothing abilities of ocean and atmospheric water. The oceans also have their Tropics to Poles heating loops that produce more uniform temperatures. The ΔK(0) of 14C at the equator corresponds exactly to the 75 W/m² difference between the input to the Tropics area and its emissions.

Water generation is helping heat to escape its home locality without any positive feedback that increases surface temperature. The return flow (not shown) would be at a lower temperature. **What would-have-been surface heat has now been transferred to atmospheric heat.** This simple explanation would backfire only if each subsequent increase in absorption (of any kind) that acts to increase the surface temperature were also to increase the average precipitable water, and thereby increase the net BBR absorption. **(That is, the runaway sequence predicted because of CO₂ would be equally applicable to runaway caused by water absorption!)**

Results in the Roper web site, a compilation of data from Wikipedia, show the (average) precipitable water as being unchanged over 60 years. There is a correlation of local water vapor with local K(0) but the percentage gains as temperature increases are lower than predicted by the Clausius-Clapeyron equation. Any correlation with CO₂ is insignificant.

It might be considered unreasonable to think that the additional 75 W/m² over the tropics range of 47 degrees in Latitude could have much of an effect on the remaining 133 degrees of Latitude, but it is *very reasonable.* Think of the earth as a round onion and start taking slices from each pole, ten degrees latitude at a time. The curved surface area on the first pair of slices is quite small, but the surface areas of subsequent rings increase as the perimeter increases. The

142   Heat convection from the Tropics

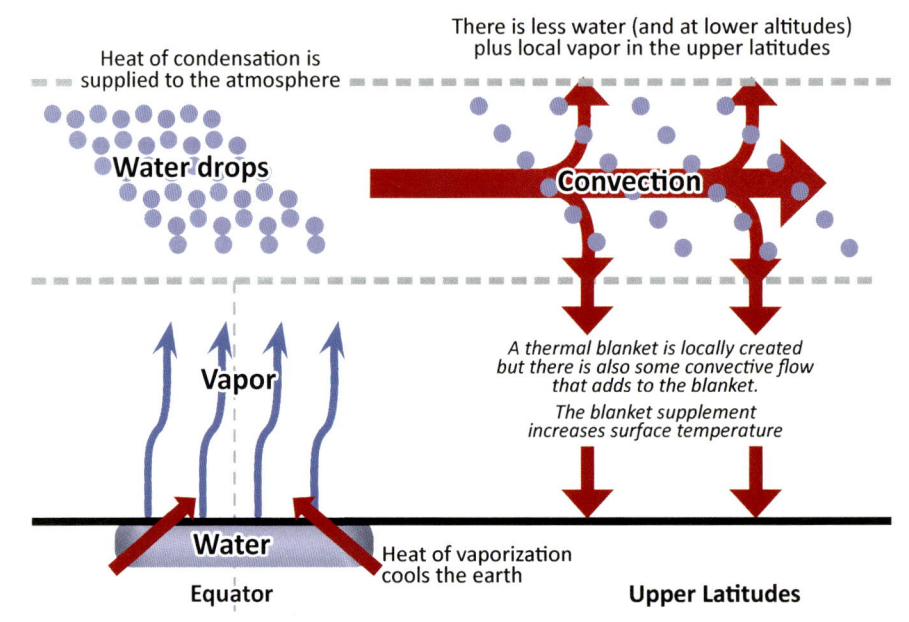

Figure 39: Convective flow of tropical conversions of vapor to water

[Diagram labels: Heat of condensation is supplied to the atmosphere; Water drops; There is less water (and at lower altitudes) plus local vapor in the upper latitudes; Convection; Vapor; A thermal blanket is locally created but there is also some convective flow that adds to the blanket. The blanket supplement increases surface temperature; Water; Heat of vaporization cools the earth; Equator; Upper Latitudes]

Author's sketch of the cooling effect of water evaporation, particularly in the Tropics, and the atmospheric condensation that produces a "heat train" flux extending towards the Poles. Blanket flux is reduced in the Tropics and increased elsewhere.

area increases are indicated in **Figure 40**.

The surface area of the Tropics is actually 40% of the area of the entire globe. The sun's heat input to the Tropics is greater than the heat input to the entire remainder of the globe. The maximum vertical spacing in **Figure 40** can be considered as being the perimeter of the equator, with the earth's "skin" having been slit along the Greenwich Median longitude and folded back. The left to right direction is a distortion of North and South Pole distances to the equator and is given in latitude. The sense of a "funnel flow" from the Tropic region towards the Poles is easily apparent. The Temperate regions represent about 51% of the area of the earth and the Polar

Global Warming Temperatures and Projections   143

Figure 40: *Relative surface areas for tropical, temperate and polar zones, with atmospheric heat flow from the Tropics.*

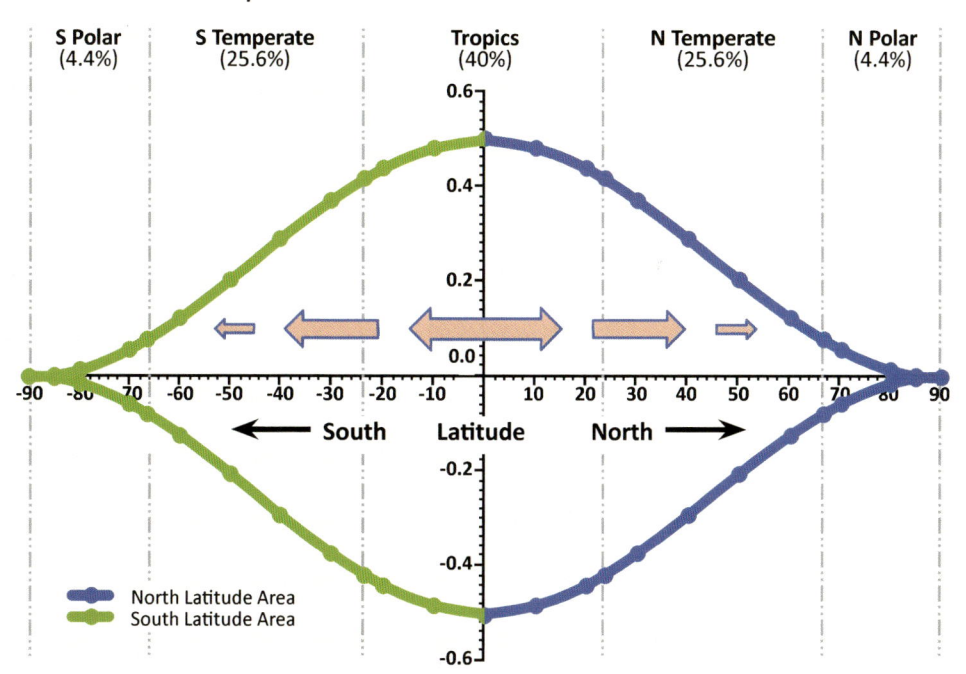

A schematic of the reduction in the earth's surface area as a function of latitude. The Tropics represent 47 degrees of the 180 degrees in latitude but comprise 40% of the earth's area. The excess heat from the Tropics, representing about 14C of possible surface temperature, easily funnels across the Temperate Zone (25.6% of the area on either side), potentially raising its average surface temperature by 5.0C.

regions only 9%. Since the Polar regions have temperatures 15C and 25C greater than their inputs from the sun, it is clear that the "funnel" (along with ocean currents, of course) clearly works out very well indeed. If half of that excess heat convected upwards and half downwards, then an "average" of 25 W/m² would be available as additional blanket. **25 W/m² corresponds to a ΔK(0) of almost 5.0C.**

   {Side comment. Almost everything about the composition of the earth appears to be ideal. The distance from the sun that keeps us warm. The Nitrogen for the soil and for absorbing the deep UV. O_2 for us to breathe. CO_2 for plants to breathe and to produce food and flowers and return O_2. The abundance of H_2O that is absolutely crucial for life, and its additional abundance that regulates the average temperatures. And H_2O's latent heat of fusion and heat of condensation that keep us from getting our immediate surroundings too cold or too hot. The fact that the Tropics occupy so much area and have such a high and almost intolerable temperature, but are, nevertheless, 14C "cooler" than their input from the sun would predict, makes one think that life might even be better if earth were farther from the sun! The northern Northern and southern Southern Temperate zones might be less livable, but would the net "desirable" living area be larger and better? It's quite possible, however — probable, in fact — that the role of atmospherically transported water and of rain would be seriously impaired. Atmospheric water is clearly dependent on the sun and on the availability of surface water, and not so much on K(0) temperature, where so much modeling is concentrated. Oceans slicing through continents would, in fact, be very desirable in reducing drought areas. Data discussed in this essay shows that, despite the assumed increases in temperature, as small as they may be, absolute humidities are not increasing and relative humidities are reducing. **The sun would extract more water from the earth if such water were available.** And local temperatures would decrease.}

A series of modifications (additional options) have been made to the original spreadsheet. The original calculations considered only CO_2, with everything else considered as Oth. For the baseline condition, Tcomb is 0.1570, *eff*TCO2 is 0.73364 and TOth is 0.214. {0.2140 • 0.73364 = 0.1570.} **So 0.214 equals *eff*TH₂O [*eff*TH₂O(Abs) • *eff*TH₂O(Conv)] times a new TOth.**

Since *eff*TH₂O includes both the absorption within the actual (average) TPW and the contribution to the blanket from evapotranspiration, *the "new" TOth background might be expected to be greater than 0.80, and certainly not be 0.2140.*

A special column for TH₂O was inserted into the spreadsheet. The first runs were done with TPW = 10 cm and with the vapor

Global Warming Temperatures and Projections 145

Figure 41: EffTH₂O from the combination of BBR absorption and the ΔBlkt(0) contribution from the convection subsequent to water vapor condensation

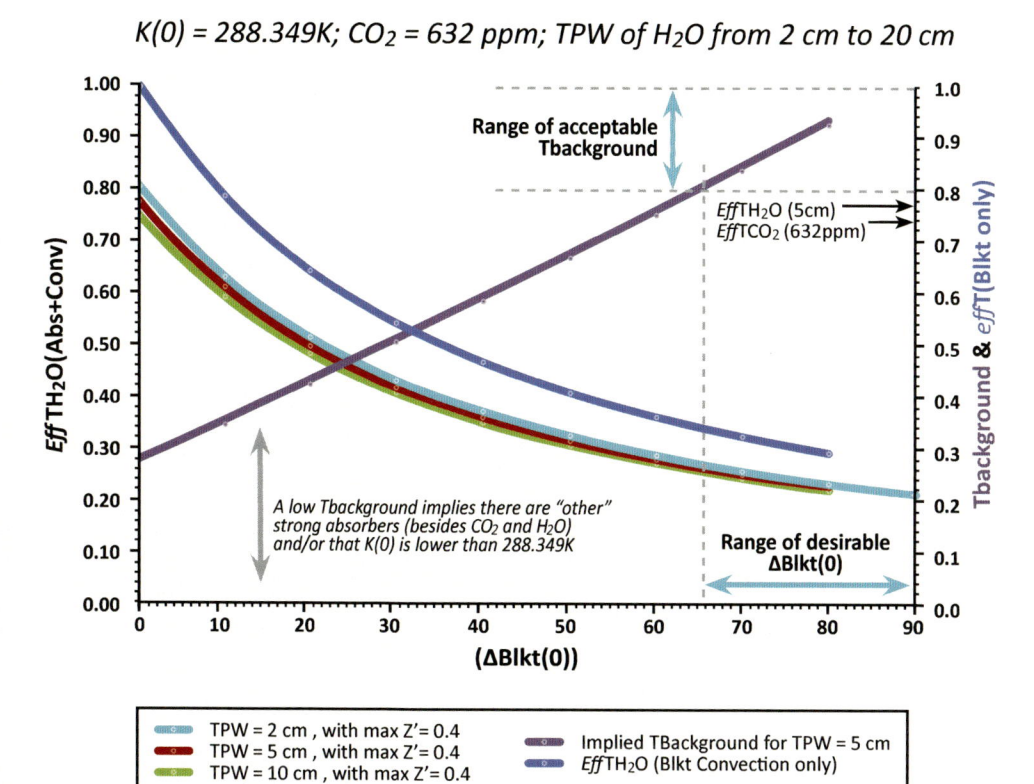

K(0) = 288.349K; CO_2 = 632 ppm; TPW of H_2O from 2 cm to 20 cm

[Chart showing EffTH₂O(Abs+Conv) on left axis (0.00 to 1.00) and Tbackground & effT(Blkt only) on right axis (0.0 to 1.0), plotted against ΔBlkt(0) from 0 to 90. Annotations include: "Range of acceptable Tbackground", "EffTH₂O (5cm)", "EffTCO₂ (632ppm)", "A low Tbackground implies there are 'other' strong absorbers (besides CO_2 and H_2O) and/or that K(0) is lower than 288.349K", "Range of desirable ΔBlkt(0)"]

Legend:
- TPW = 2 cm, with max Z' = 0.4
- TPW = 5 cm, with max Z' = 0.4
- TPW = 10 cm, with max Z' = 0.4
- Implied TBackground for TPW = 5 cm
- EffTH₂O (Blkt Convection only)

Plots of effective T (transmissivity) as evapotranspiration makes more contributions to the Blanket. The left axis refers to *eff*TH₂O for the combined BBR absorption by water and the condensation of water vapor within a simultaneous concentration of 632 ppm of CO_2. Three values of TPW (2, 5, and 10 cm) are considered. The right axis refers to *eff*TH₂O considered only for the Blanket; the axis also indicates the discrete values for *eff*TH₂O(BBR only) and *eff*TCO₂; it also indicates the remainder TOth = Tbackground that must exist if the temperature is to be an equilibrium temperature for each blanket value.

density constant in the Z' (altitude) direction until its cutoff at Z' = 0.4 (Z = 3.91 km). {That is, the integral of water vapor from Z' = 0 to Z' = 0.4 equals 10 cm.} The Stull et al tables provide the individual entries for TH_2O (WN,10 cm). The spreadsheet slices provide the appropriate values for TH_2O within each slice, with TH_2O = 1.0 beyond Z' = 0.4. CO_2 was set equal to 632 ppm and K(0) = 288.349, the known values from the earlier calculations. There is a separate column for TCO_2 (WN,632 ppm), a third column for the unknown constant value TOth and a Tcomb column for Tcomb = TH_2O(WN) • TCO_2(WN) • TOth. Since (1-Tcomb) is precisely related to K(0) via the calibration plot (**Figure 13**) the formula for *eff*TH_2O for X cm of H_2O is:

$$\textit{eff}TH_2O(X\ cmTPW,\Delta Blkt) = \left[\frac{1-(1-Tcomb)}{\textit{eff}TCO_2 \bullet TOth(X\ cmTPW,\Delta Blkt(0))}\right]$$

ΔBlkt(0) is the (average) free drop of Blanket associated with the Tropic convection plus the average evapotranspiration within the local area that also returns as a thermal blanket. The average value for ΔBlkt(0) that is anticipated from arguments yet to be explained is about 70 to 90 W/m². So if an average of 25 W/m² were to be supplied by the Tropics, then 45 to 65 would be provided "locally."

The simulation process is identical to those previously described, but ΔBlkt(0) is also an input. For example:

set ΔBlkt(0) = 70 W/m², CO_2 = 632 ppm, K(0) = 288.349K, and TPW = 10 cm.

The required Blkt(0) = 152.97 for K(0) = 288.349K, and TOth is adjusted for each ΔBlkt(0) until this Blkt(0) is obtained.

For ΔBlkt(0) = 70 the result is **(new) TOth = 0.8614**. From this we can determine the *eff*TH_2O. 0.2140 = 0.8614 • *eff*TH_2O(10 cm, 70). So ***eff*TH_2O(10 cm, 70) = 0.2484.**

The (1-T) absorption ratio of H_2O to CO_2 is 2.8.

This absorption ratio is high, but, perhaps, not as high as what one might expect for TPW = 10 cm. However 45% of Blkt(0) is being supplied by the convection associated with evapotranspiration. The new TOth = Tbackground of 0.8614 is marginally okay, because,

Global Warming Temperatures and Projections

Figure 42: K(0) Dependence of EffTH$_2$O from the combination of BBR absorption, evapotranspiration, and water vapor condensation in the atmosphere, with a convection contribution of ΔBlkt(0) to the Blanket.

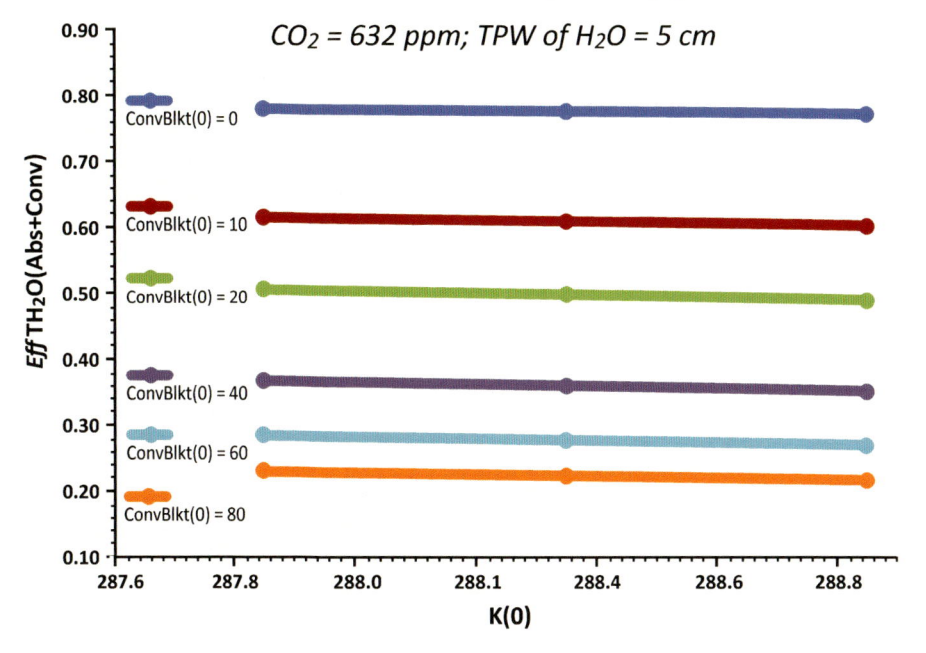

EffTH$_2$O(Abs+Conv) vs. K(0), with additional components of Blanket associated with evapotranspiration as parameter.

after all, *eff*TCO$_2$ is 0.73364.

This simulation was followed by a sequence of simulations with additional "free" Blanket entries. (The simulation, after all, is meant to represent the "average" state.} The ΔBlkt(0) values were 0 W/m² (for which T0th = 0.2850, an unacceptably low transmission for an "unknown" background), 10, 20, ..., 70, 80 W/m². **Tables 6** and 7 summarize the results for TPW = 10 cm and TPW = 5 cm. **Figures 41** and **42** provide graphical summaries.

*Eff*TH$_2$O without any contributions from evapotranspiration is only 0.75086. This, along with the *eff*TCO$_2$ = 0.73364, is NOT enough

Table 6: EffTH$_2$O for TPW = 10 cm and CO$_2$ = 632 ppm

Water (up to Z'=0.4) (cm)	Blkt Addition (W/m²)	effTH$_2$O (Conv +Abs)	TOth (remaining bkgnd)	effTH$_2$O (Conv only)	(1-TH$_2$O) / (1-TCO$_2$)
10	0	0.7509	0.2850	1.000	0.935
10	10	0.5929	0.3609	0.7897	1.528
10	20	0.4870	0.4395	0.6485	1.926
10	30	0.4115	0.5201	0.5480	2.209
10	40	0.3550	0.6028	0.4728	2.421
10	50	0.3114	0.6873	0.4147	2.585
10	60	0.2766	0.7736	0.3684	2.716
10	70	0.2484	0.8614	0.3309	2.822
10	80	0.2251	0.9507	0.2998	2.909

EffTH$_2$O and TOth for K(0) = 288.349K and CO$_2$ = 632 ppm as Blkt additions from evapotranspiration are added. **TPW = 10 cm** = 42,600 ppm$_v$. The 10cm total of water is distributed with a constant density on the Z' scale from Z'=0 to Z'=0.4. Consistent parameters are: effTCO$_2$ = 0.73364; effTH$_2$O (from Abs) = 0.75086; and (1-Tcomb) = 0.8424. {The calibration plot links (1-Tcomb) with K(0).}

to provide the (1-Tcomb) value of 0.8424 necessary to assure a K(0) of 288.349K unless there is a substantial and unrealistic level of background absorption. The TPW of 10 cm of water is no more effective than 632 ppm of CO$_2$ (last column). Blanket additions from convection of 70 to 90 W/m² are necessary in order to remove the label of "unrealistic" from the background absorption.

Water is exhibiting the same tendency towards limitations in its absorption as its concentrations are increased as has been observed for CO$_2$. The set of values for effTCO$_2$ at {200, 500, 1,000} atm-cm (which is the same as {253, 632, 1,264} ppm) is {0.7706,

Global Warming Temperatures and Projections   149

Table 7: EffTH$_2$O for TPW = 5 cm and CO$_2$ = 632 ppm

Water (up to Z'=0.4) (cm)	Blkt Addition (W/m²)	effTH$_2$O (Conv +Abs)	TOth (remaining bkgnd)	effTH$_2$O (Conv only)	(1-TH$_2$O) / (1-TCO$_2$)
5	0	**0.7760**	0.2758	1.000	**0.8410**
5	10	0.6116	0.3499	0.7882	1.458
5	20	0.5017	0.4266	0.6465	1.871
5	30	0.4233	0.5055	0.5456	2.165
5	40	0.3649	0.5864	0.4703	2.384
5	50	0.3198	0.6692	0.4121	2.554
5	60	0.2839	0.7537	0.3659	2.688
5	70	**0.2549**	**0.8397**	0.3284	2.798
5	80	0.2308	0.9272	0.2974	2.888

*Eff*TH$_2$O and TOth for K(0) = 288.349K and CO$_2$ = 632 ppm as Blkt additions from evapotranspiration are added. **TPW = 5 cm = 21,300 ppm$_v$.** Consistent parameters are: *eff*TCO$_2$ = 0.73364; *eff*TH$_2$O (from Abs) = **0.7760**; and (1-Tcomb) = 0.8424. {The calibration plot links (1-Tcomb) with K(0).}

0.7336, 0.7005}. The set of values for *eff*TH$_2$O for {2, 5, 10} cm TPW is {0.8043, 0.7760, 0.7509}. The overall effect of water on creating the blanket does not become significantly greater than that for CO$_2$ until the additional Blanket from evapotranspiration and convection exceeds 20 W/m². With additional Blanket values exceeding 50 W/m², the "absorption" ratio of water to CO$_2$, as measured by (1-TH$_2$O)/(1-TCO$_2$) exceeds 2.5 (last column).

There is no absolute determination that can be made (within the spreadsheet) of what the average contribution from the Tropic and local sources must be. **It appears that a contribution in the range**

of 70 to 90 W/m² would be suitable, with something more going to latitudes lower than the average and something less to latitudes higher than the average. Such a contribution also appears to be realistically probable. It may not just be fortuitous that the range of 70 to 90 indicated by the Tables agrees perfectly with the Trenberth et al value of 80 W/m² for evapotranspiration in their average model.

In **Figure 41**, the uppermost line (in blue) gives the *eff*TH$_2$O **considered only for the Blanket.** The triplet of plots below this gives the result when the absorption for three concentrations of TPW are included. (Yes, one might expect the increase in evapotranspiration and the increase in TPW to be mutually dependent, but the local evapotranspiration and local BBR absorption are only poorly related to the convection component from a distance.) Since the triplet is really a quadruplet, it is the *eff*TH$_2$O(Blkt) that is forcing the "quadruplet" towards lower *eff*T (and higher effective absorption). Some Tbackground is also needed to push the final Tcomb to 0.1570, the approximate value for a K(0) of 288.349K.

Since K(0) is constant for every point on this plot, then Tcomb must not deviate from 0.1570. The multiplication, e.g., of *eff*TCO$_2$ • *eff*TH$_2$O(Abs+conv, 5cm) • Tbackground (TPW = 5cm) must equal 0.1570. The actual product for ΔBlkt(0) = 20 is

0.73364 • 0.5017 • 0.4266 = 0.1570

and for ΔBlkt(0) = 60 the product is:

0.73364 • 0.2839 • 0.7539 = 0.1570.

If, for example, Tbackground for ΔBlkt(0) = 60 were 0.9272, Tcomb would equal 0.1931, and K(0) would be close to 260K!

Once again CO$_2$ is seen to be a rather insignificant player within a scenario of so many other powerful and variable influences. **If perturbations could produce a runaway, a runaway would have already occurred.**

Figure 42 shows that the ultimate value of *eff*TH$_2$O is only weakly dependent on K(0). That is, small average changes in Blanket will not significantly impact K(0). This has already been firmly established. It takes about 10.8 W/m² of atmospheric BBR absorption to produce the 5.4 W/m² of Blkt(0) that will increase K(0) by 1.0K.

Water as a 'climate changer,' with replacement of the Tropical heat bus by circulation cells

This document has concentrated on global warming and on validity of theories that humans are producing an inexorable increase in global temperature. All utilizations of power result in the ultimate creation of heat. The daily power of human fossil activity is at least ten times lower than the *standard deviation* of the sun's insolation. A single, extended, natural El Niño event can, however, change the temperature of a hemisphere for 2 to 3 years, and the ocean water temperature changes totally disrupt the atmospheric CO_2 levels. The earth's contribution to the energy it takes to produce the water elevation (independent of consideration of rotational kinetic energies!) for a *single* three to five day hurricane is close to all the energy being created by all humans during that time. Humans can tap into nature's energy, and we can partially protect ourselves against its awesome power, but we cannot overcome its power. (See **Quick references (5 tables)** after the appendicies.)

Ocean water and ocean currents modify peaks and valleys in

152   *Water as a 'climate changer'*

temperature, and, as explained above, water evaporation from the surface of the earth cools the earth and produces a heat exchange with the atmosphere. An atmospheric heat bus can then carry heat towards the poles. These are all much more significant events than absorptions by CO_2. If this heat bus were not in existence, then, perhaps, some might be clamoring for intentional CO_2 releases to raise the Temperate zone temperatures by the 5.0 or so Centigrade that would *not* be being achieved by non-existent Tropics to Polar buses.

There are many scientific rejections to the concept of having uniform weather with little to no "climate change" disturbances. Human disrupters produce puny changes; nature produces megafold greater disruptions. They include more than just seasonal changes, earthquakes, volcanoes, and thunderstorms. Nights are not long enough to allow any return to the same equilibrium each morning; temperatures can change by 10 to 20C just overnight (that's 20C, not the scary, runaway 2.0C we hear about). Water and its circulation is a random disrupter.

The Tropical heat bus, presented in **Figure 39**, is too simplistic. Tropical heat is associated with hot air winds projected from about 24°L toward the Poles and initially at relatively high altitudes. See **Figure 43**. At the same time Polar cold winds above 60L are projected towards lower Latitudes, initially at relatively low altitudes. Rotation of the earth produces curvatures of the S to N winds to the east and N to S winds to the west, but the net directions are either to the north or the south. A crash of moisture-bearing winds, which occurs at a *low pressure* around 60°L, is quite dramatic and produces precipitation. There is also a clash at *high pressure* around 30°L. The hot upper winds from within the tropical zone have already been denuded of their moisture by the heavy rains within the tropical zone and have little moisture to contribute to this crash with the cool upper winds above 30°L projected towards the equator.

Figure 43 shows three *high to low* altitude circulation cells within each hemisphere. Each cell is actually a "tube" that encompasses the earth and whose precise local location can shift with air pressure. The rising hot air at the equator produces a "Low" in pressure. At about 30°L this air is still hot but has cooled sufficiently such that its densification produces a sinking column and a local

Global Warming Temperatures and Projections

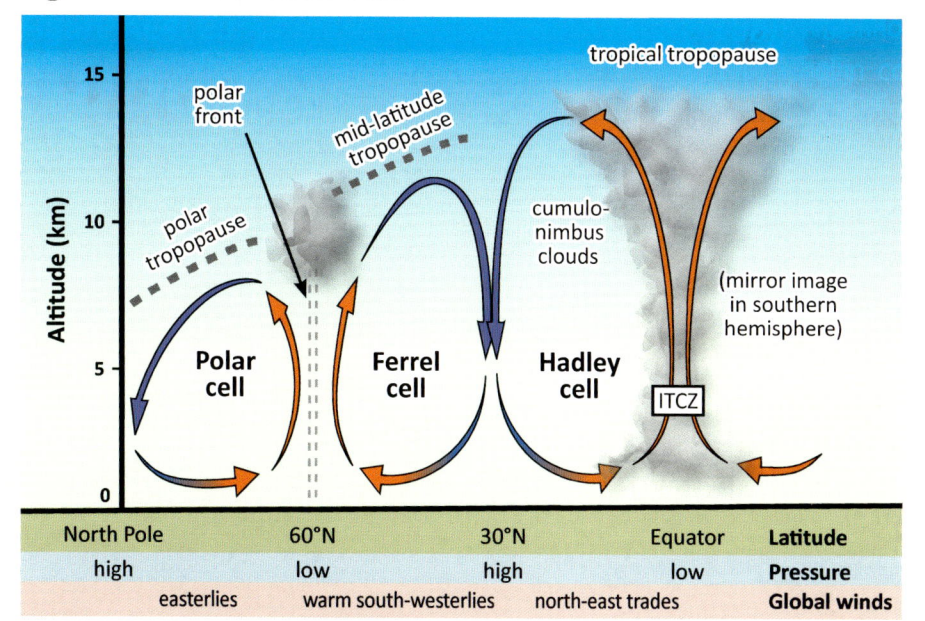

Figure 43: Circulation cells

[Diagram showing atmospheric circulation cells with the following labels:
- Altitude (km) axis from 0 to 15
- tropical tropopause
- polar front
- mid-latitude tropopause
- polar tropopause
- cumulo-nimbus clouds
- Polar cell, Ferrel cell, Hadley cell
- ITCZ
- (mirror image in southern hemisphere)

Bottom table:
| North Pole | 60°N | 30°N | Equator | Latitude |
| high | low | high | low | Pressure |
| easterlies | warm south-westerlies | north-east trades | | Global winds |]

A Cross section of the three circulation cells (globe-circling tubes) that exist in each hemisphere. The tropopause altitude is the altitude above which turbulences are significantly reduced. The Ferrel cell acts as the coupling cell that attempts to regulate the hot blasts from the tropics with the cold blasts from the Poles. The indicated 30°Lat and 60°Lat boundaries fluctuate, and, particularly so, the 60°N boundary. Significant rainfalls occur at the pressure Lows at the equator and 60°Lat.

"High" pressure. This rather consistent "High" and "dry" at 30°L is associated with deserts around the world. This ideal fully looped cell/tube between 0° and 30°L is the Hadley cell. It's rather strange to have a desert at the outer edge of a Tropical Zone, especially since the rainfall within the Tropical zone is intense. Phoenix AZ weather is a good example.

Those two sinking columns of air at the edges of the "Hadley" and "Ferrel" cells are, not, however, maintaining their own lanes, but are intimately mixing. (The mixing may spread well into the

Ferrel cell, nearly obliterating any clear boundaries.) There is warm air that is left for latitudes greater than 30°L, but reduced moisture. The lower level air returning to the Tropics in the Hadley cell is cooler than it would be if the loops had not been mixing; the ultimate effect is to reduce the surface temperature in the Tropics by 14C. The warmer air in the Ferrel cell that is returning towards the Pole is a correction to the single cell "bus" flow in **Figure 39**.

The cell/tube from 30°L to a variable 60°L is the Ferrel cell. The clockwise (as shown) rotation of the flow from the High at 30°N to a warmer, water-containing Low at 60°N rises at 60°N along with the rise of the CCW Polar cell. The two-lane crash of cold air with warm water-containing air produces a very significant precipitation in the region of 50°N to 70°N.

The rotational kinetic energy of the earth (West to East; CCW when observed from above the North Pole) and temperature-influenced changes in atmospheric pressure are the sources of power for most storms. Earth surface speeds are greater than 1,000 miles per hour at the equator and drop to 0 at the Poles. The set of three circulation cells — as opposed to other options of one, five, etc. — must be the most energy efficient mode of nature's possibilities. The rotational vector of the earth (directed along the pole axis), and the local rotational vectors of the circulation cells, are mutually perpendicular, and, along with pressure changes, this sets the framework for predictable but complicated weather patterns. The Polar cell vertical flow matches up against the Ferrel cell, and the Hadley cell vertical flow matches up against the Ferrel cell. Nature is further satisfied by having the axes of the circulation cells aligned with the local surface winds (easterlies, westerlies, and trade winds).

The degrees for match-up values are not rigid, and of such clashes are Jet streams and "Polar Vortexes" formed. Nature doesn't care what humans are doing, just as it doesn't care when earthquakes strike.

Figure 44 is the support for the preceding discussion. **Figure 44** is from the website www.roperld.com/science/PrecipLatitude_Longitude.htm. It presents the 60-year pattern of average precipitation rates in mm/day vs. Latitude.

This plot is not as quiet as the 60-year averaged temperature

Global Warming Temperatures and Projections

Figure 44: *The averaged precipitation rate in mm/day vs. Latitude over a 60 year span (at ten year intervals)*

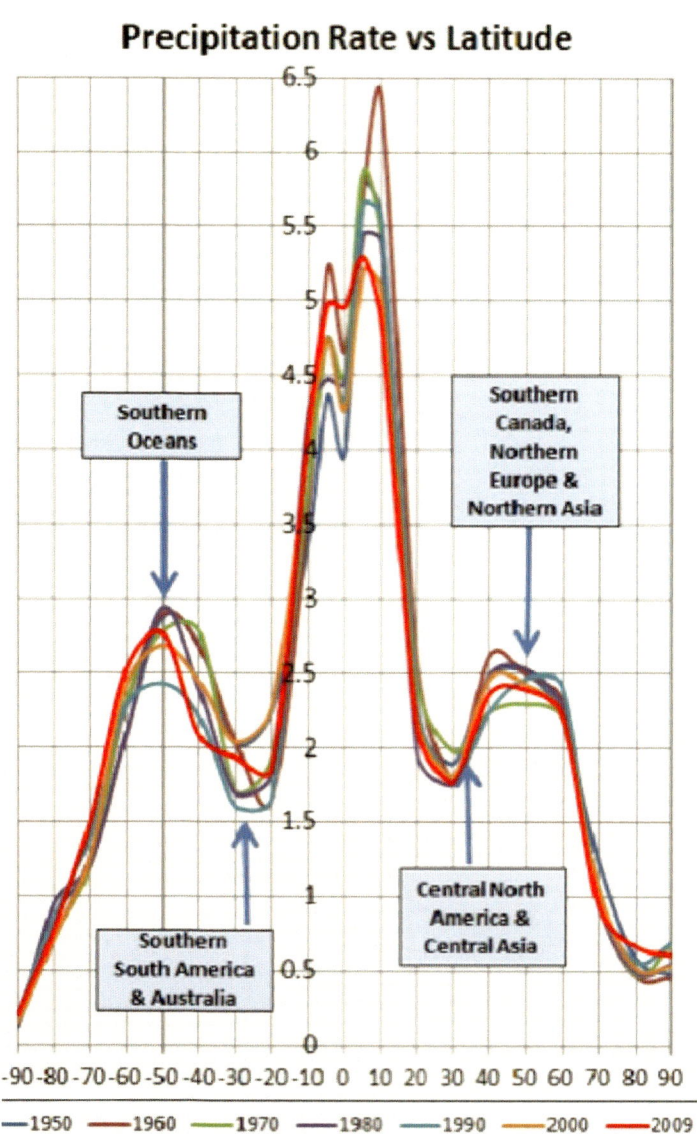

Precipitation Rate vs Latitude

Credit: www.roperld.com/science/PrecipLatitude_Longitude.htm

The effects of the interactions at the boundary of the Hedley and Ferrel and the boundary of the Polar and Ferrel cell are apparent. Temperature may decrease monotonically as Latitude increases, but precipitation has its own pattern.

plot, but it definitively shows the kind of natural climate changes that must be expected. Note, e.g., the "desert dip." The air has already been fairly depleted of water by the Tropical rains, **but the sun keeps evaporating any water it can find. So the desert gets even dryer and maintains a low RH.**

Variations — even changes in the "count" of hurricanes — cannot be blamed on humans. The L. D. Roper data is the averaged data over an entire month for the median month of that year. There is no standard deviation presented for that month, nor for an entire year. **It is the relative consistency of the average within a highly turbulent natural process that is important.**

The plot appears to demonstrate another gift to us from nature. It would appear from this plot that without the "clashing" of circulation cells there might be very little rainfall for Latitudes beyond 40 degrees. Nature, however, is not trying to create as much rain as possible, since nature tries to have optimum overall energy and power efficiencies within all its options. With a single cell (per hemisphere) the sum of the kinetic energies might be higher and the surges might be more intense, and wasteful. The circulation tubes are the equivalents of the natural "vibration modes" that are seen throughout all of nature. These modes act as energy "storage" sites and they have their own pecking order of relative importance. The sharing of the "crown" of heat from the Tropics is surely a blessing to the rest of the globe. The spinning of the earth at latitude-dependent speeds, the immense rotational kinetic energy associated with that spinning, the clouds, the daily sun cycles, and the Coriolis effect all provide the turbulence that enhances the variety of our climate. The higher latitudes appear to have both more heat (higher temperature) and more water than they would without the circulation cells, but no one considers that a fatal hazard to human existence.

Section Two summary

The discussions and contributions within this chapter on water appear to be more valuable in understanding global temperature increases than the solo consideration of CO_2 in the major portions of the book. The solo consideration of CO_2 serves its purpose in demonstrating that CO_2 is an insignificant player either in any global warming or in any possible meltdown or runaway effects. The universal calibration plot is of great value in assessing the mutual interactions of multiple absorbers. The total contribution of 400 ppm of CO_2 to the blanket is well below 10%, AND even a quadrupling of CO_2 will add only 1% to the final blanket.

The calibration plot has also been employed with great advantage within this chapter on the effects of water vapor. A given K(0) can always be associated with a particular absorption/feedback. (For example, Tcomb = 0.157 corresponds to a K(0) of 288.35K; Tcomb = 0.193 corresponds to a K(0) of 260K.)

Although there are scientific arguments that water and CO_2 have an intimate, although once-removed, interaction that greatly enhances global warming, those arguments were challenged within the CO_2 discussions. The premise has been that increased CO_2 raises the temperature (even if hardly at all), that this increase in temperature enhances water evaporation (and that CO_2 must be blamed), that this water evaporation has a residence time in the atmosphere that increases BBR absorption by its molecules, that there is an amplifying effect that increases the blanket flux and raises the temperature until some self-limit (if lucky) occurs or runaway (if not lucky) develops. The truth is that the first result of water evaporation is a cooling of the earth. But, yes, each evaporated H_2O molecule, as long as it remains in the atmosphere, acts as an absorber, and each absorption will have half its absorbed energy returned within the blanket.

The heat extracted from the earth in order to produce water vapor is not simply lost. It is released higher up in the atmosphere when the vapor molecules condense to water droplets. If all this heat worked its way back to the earth's surface, the cooling effect on the earth would be nulled. If much of this heat convects laterally towards cooler temperatures of the globe, as is definitively the case with Tropics heat, then the cooled surface remains cooled and higher latitudes become warmer. This is clearly a double benefit. The surface of the tropics is 14 degrees cooler and upper latitudes experience comfortably (hopefully) higher temperatures.

There is no verification that a higher temperature necessarily produces a higher water density. The equilibrium vapor pressure is increased but temperature-induced evaporation is dependent on the availability of water and the equilibrium vapor pressure. Final surface temperatures can, in fact, be expected to decrease because of evaporation. The sun has a close-to-instantaneous effect on producing evaporation, and at the equator there is plenty of sun and water. The earth in the Tropics is 14C cooler than it might be expected to be. Deserts are hot not just because of the sun, but because there is no earth water left that can cool the desert by means of evaporation. Temperatures drop on either side of a desert.

Excess temperature will be found anywhere in pre-desert climates where there is very little surface water. Plots show that RH, except for over the oceans, has been decreasing with time. This may be a real climate changer, much more so than CO_2. Are ground surfaces losing water "naturally" with time? Is agriculture removing more surface water than it is adding?

At the equator there is a considerable amount of evaporation, a large TPW, and plentiful rainfall. But the rainfall does not return, as warm rain drops, nor does convection return, all of the heat that was extracted at the time of evaporation.

Earlier figures described the surplus heat over the Tropics. All of that surplus represents the 75 W/m^2 (14C) reduction associated with K(0). Other figures show how this heat convectively spreads into the Temperate and Polar regions, both downwards and upwards. This is not too difficult since the Tropics collection area is 40% of the entire surface area and the Pole-ward convections are traversing smaller and smaller inter-latitude areas. This convection becomes a component of the blankets at each latitude, and this, of course, increases surface temperatures. An average increase of 25 to 30 W/m^2 of blanket from beyond local latitudes increases temperate zone K(0) temperatures by about 5C.

This document can provide no direct analysis that determines what this "average" increase in blanket as a function of Latitude actually is, but the estimations are quite good. They are "good" in the sense that the averaged modeling presented in this document needed an unspecified Background blanket of about 80 to 90 W/m^2 in order for its verification of baseline conditions {CO_2= 380 ppm, K(0) = 288.0K) to be valid. This agrees with the evapotranspiration component from Trenberth et al. So the analysis of this section on water is quite supportive.

The three circulation cells/tubes on each side of the equator are a naturally produced effect since nature always seeks the most energy and power efficient solutions. A single cell could theoretically work, but it would be totally unstable. A two cell system would not work because a Polar mid-latitude rise would totally clash with a Tropical mid-latitude sink. The easterlies, westerlies, and trade wind directions are all aligned with the rotational vectors of their respective cells/tubes. The earth's surface speeds in directing these hot and cold winds are the impetus for creating the actual rotations within the circulation cells.

The blanket contribution by water evaporation and condensation is slightly greater than the blanket contributions by BBR absorptions by the atmospheric molecules. The combination of the section on CO_2 and the section on water are totally compatible with the Trenberth et al results, show the validity

of the Stull et al tables, and provide valid projections whenever the fundamental input parameters are changed.

Climate situations such as those in the summertime desert of Arizona provide an additional avenue for the introduction of heat, an avenue that is the dual of the convection from the Tropics. If previous conclusions and figures in this document are accepted, then the following comments present concepts for future study. The hottest days in Phoenix have highs of about 120°F (49°C) and lows of about 80°F (27°C). Overnight is much too short a time for a near-equilibrium to be achieved and the Tropics cannot be called upon for additional convection. Consider a period of many such days; a repeating daily "equilibrium" must necessarily result. Heat that would like to escape is still stored in the earth and atmosphere by daybreak. This storage keeps building up until it can, over an entire day, supply (and store) the additional necessary W/m^2 requirements. **The average temperature (100°F), when applied to an extension of the calibration plot of Figure 13, correlates with a Tcomb value of ~0.01. The difficulties, and dangers, of having Tcomb values approaching 0.0 have already been pointed out, and it would appear that nature adapts by increasing earth temperatures so as to achieve a larger energy storage.**

The average daily solar input in Phoenix in June and July appears to be about 360 W/m^2, considerably more than the annual global average of 239 W/m^2 employed throughout this document. With some reasonable assumptions, the calculation procedures of this document conclude that the *eff*TConv must be about 0.023. Convection plus water absorption (and there surely will be some) must be about 93 W/m^2, and P(0) minus the solar average is 169 W/m^2, which is the total required Blanket. **Were the solar average to be reduced to 300 W/m^2, with the average K(0) maintained, then the required Blanket is a predictable 229 W/m^2. Any increased "convection" must be coming from heat storage within the earth.** The author hopes that someone will pursue this issue of deserts, with the goal of predicting what the realistic maximum of K(0) on earth is likely to be. It is expected that it will be about 140°F (the maximum currently recorded temperatures over successive days?). This effect is solely a solar effect, not a CO_2 effect.

Section Two references

Published distributions of water vapor:
- The Stull et al reference ("The Infrared Transmittance of Water Vapor," by P. Wyatt, V. Stull & G. Plass. Applied Optics, vol. 3, No. 2, Feb 1964) has WN-dependent absorption tables for 2, 5, 10, ... centimeters of precipitable water.
- K(0) calculations, with and without CO_2, are discussed below.
  - Clive Best (clivebest.com/blog/?p=4517) provides several published plots of precipitable water vapor: http://www.leif.org/EOS/2012GL052094-pip.pdf "Weather and Climate Analyses Using Improved Global Water Vapor Observations"
  - Climate Science Change Essay, https://friendsofscience.org/index.php?id=681 and www.friendsofscience.org/index.php?id=710
- Another highly recommended reference is: www.roperld.com/science/PrecipLatitude_Longitude.htm. It provides overlays of 60 years of plots (seven at ten-year intervals) of:
  - Precipitation rate vs. Latitude;
  - Precipitable water (kg/m^2) vs. Latitude;
  - Temperature vs. Latitude
  - Professor Richard Seager's site, http://ocp.ldeo.columbia.edu/res/div/ocp/pub/seager/Kang_Seager_subm.pdf, summarizes the various effects that both contribute to, and detract from, the northern hemisphere having a distinctly warmer temperature offset.

Appendicies

Appendix A: Correlations with total absorption

Blkt(0) vs. Cum(Abs/2)

Figure AA1 plots **Blkt(0) vs. Cum(Abs/2) with Toth fixed at 0.214 and CO_2 varying from 0 to 12,640 ppm. K(0) is going from 284.722 to 291.155K**

With TOth fixed at 0.214, and CO_2 increasing from 0 to 12,640 ppm, Cum(Abs/2) increases from 99.55 to 133.62 W/m² and Blkt(0) increases from 133.62 to 168.45 W/m². {ΔBlkt(0)/ΔK(0)= ΔP(0)/ΔK(0) = 5.4 W/m²/K}

However, with **CO_2 = 0 and TOth reduced to 0.1175** for K(0)= 291.155, Cum(Abs/2) is 127.67 W/m². The additional contribution from CO_2 for that same K(0) is only 6.09 W/m², or 3.6% of the total Cum(Abs/2). (Blkt(0) is the same for each side-by-side pairing since K(0) is unchanged.)

The magnification, the ratio of Blkt(0) to Cum(Abs/2), decreases from 1.34 to 1.26 as CO_2 is increased. (The baseline value of about 1.3 is indicated.)

Many more photons with sensitivity to CO_2 absorption are absorbed at low altitudes, and so fewer of these particular photons

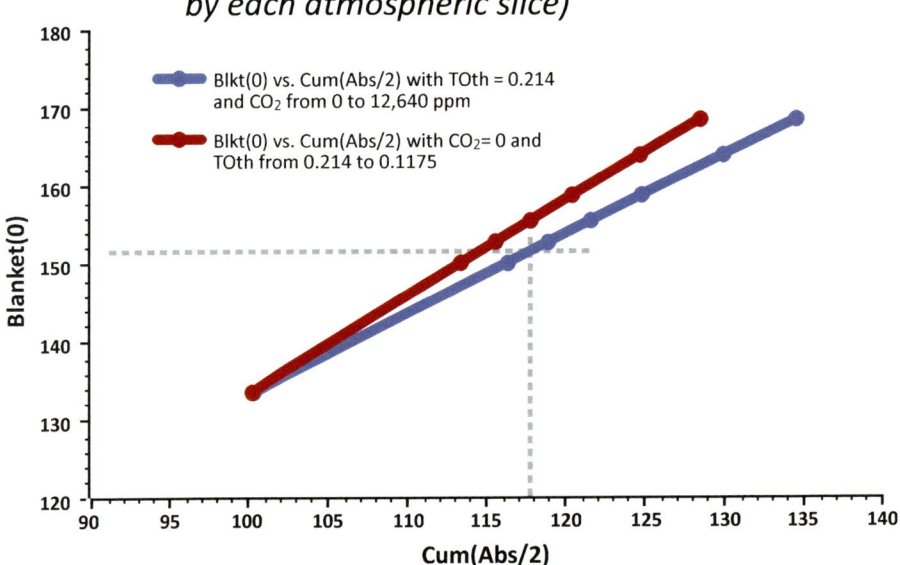

Figure AA1: *Sum of downward-directed absorption of BBR by each atmospheric slice)*

Legend:
- Blkt(0) vs. Cum(Abs/2) with TOth = 0.214 and CO_2 from 0 to 12,640 ppm
- Blkt(0) vs. Cum(Abs/2) with CO_2 = 0 and TOth from 0.214 to 0.1175

(Y-axis: Blanket(0); X-axis: Cum(Abs/2))

are "lost" to the adiabatic effect. The Blanket will necessarily increase with more CO_2, but the magnification is reduced. This is what we see happening.

{There are plots available that show the "notching" (eating away of the original BBR Planck emission) as a function of WN and altitude, but they are not presented here.}

Cum(-adiab) vs. Cum(Abs/2)

Cum(-adiab) is the sum of the pre-absorption adiabatic losses for the BBR within each slice. {Cum(adiab) has a negative value; Cum(-adiab) has a positive value.}

Cum(Abs/2) is the sum of the actual absorption losses in each slice, divided by 2 to represent the post-absorption downward emission. After magnification by the reverse (heating) adiabatic effect, K(0), P(0), and Blkt(0) are established.

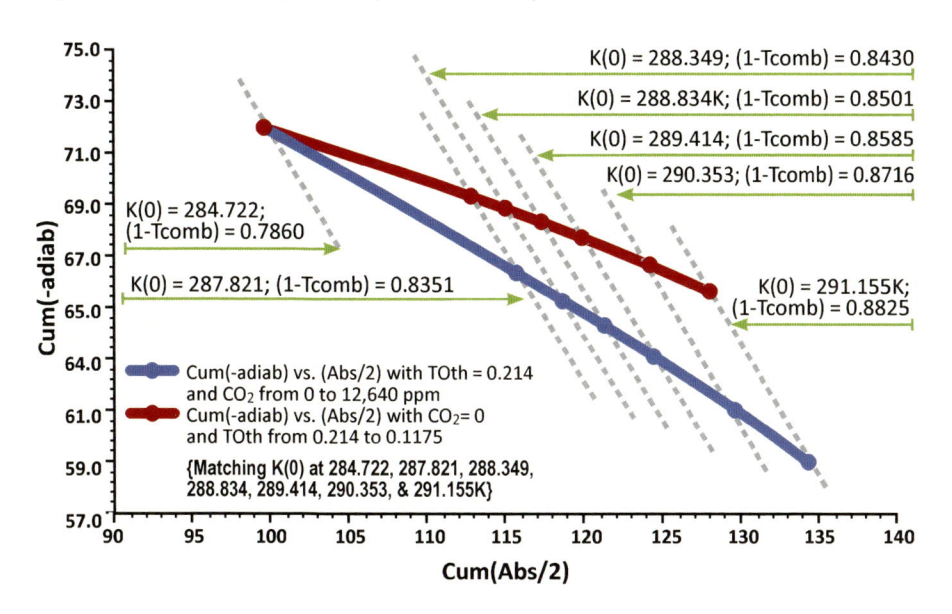

Figure AA2: Z(km) vs Z' (nomalized)

Plot showing Cum(-adiab) vs Cum(Abs/2):
- Blue curve: Cum(-adiab) vs. (Abs/2) with TOth = 0.214 and CO_2 from 0 to 12,640 ppm
- Red curve: Cum(-adiab) vs. (Abs/2) with CO_2 = 0 and TOth from 0.214 to 0.1175
- {Matching K(0) at 284.722, 287.821, 288.349, 288.834, 289.414, 290.353, & 291.155K}

Labels on plot:
- K(0) = 284.722; (1-Tcomb) = 0.7860
- K(0) = 287.821; (1-Tcomb) = 0.8351
- K(0) = 288.349; (1-Tcomb) = 0.8430
- K(0) = 288.834K; (1-Tcomb) = 0.8501
- K(0) = 289.414; (1-Tcomb) = 0.8585
- K(0) = 290.353; (1-Tcomb) = 0.8716
- K(0) = 291.155K; (1-Tcomb) = 0.8825

Except for whatever BBR can be considered an "escape" above a specified altitude, all losses end up being adiabatic cooling losses. The post-absorption emissions *reintroduce* all absorbed energy.

In effect, *half of the absorbed flux is inhibiting the opportunity for adiabatic loss*. The surface temperature rises and this improves the opportunity for adiabatic loss. This is a positive feedback effect, but it has an equilibrium and it is not a runaway effect. CO_2 has no special role that it plays.

The more absorption there is, the larger is the blanket, the higher is the average temperature of each slice, and the higher is the temperature difference across each slice. Pre-absorption adiabatic losses must still occur within each slice, but this loss will be less than if there were no absorption and if the BBR were attempting to transit the entire slice width.

Figure AA2 plots Cum(-adiab) vs. Cum(Abs/2). Cum(Abs/2) is always greater than Cum(-adiab) and is increasing significantly, and Cum (-adiab) is *decreasing* at a slower rate than Cum(Abs/2) is increasing.

As the temperature increases by 6.433C and (1-Tcomb) increases by 0.0965 (0.67C per 0.01 (1-Tcomb)), Cum(Abs/2) increases by 34.21 (5.32 W/m² per 1.0C), P(0) increases by 34.83 W/m² (5.41 W/m² per 1.0C), and Cum(-adiab) *decreases* by 12.87 (-2.00 W/m² per 1.0C).

That apparent loss of 2.00 W/m² is totally restored by the complementary gain of 2.00 by the adiabatic heating effect.

The increase in P(0) is always matched by an increase in Blkt(0). (This, of course, was always known, but the breakout numbers also prove it to be so.)

Since the increase in Cum(Abs/2) is 5.3 W/m² for each 1.0C rise in K(0), the **increase in *total post-adiabatic absorption* (Cum(Abs)) is about 10.6 W/m² per 1.0C. So, alternatively, it can be said that each one degree rise in K(0) requires 10.6 W/m² of absorption** (half of which will be directed upward and will suffer adiabatic loss).

The possibility of a "runaway" depends significantly on the value of the "forcing function." Technically, the net of all forcing functions should be less than ΔP(0) per 1.0C for there to be a runaway. In this analysis it takes an absorption of 10.6 W/m² to obtain a temperature rise of 5.41 W/m² of P(0) associated with a 1.0 increase in K(0). There is absolutely no runaway no matter what the absorber is.

Appendix B: Normalized altitude Z'

The advantages of Z' (normalized altitude)

Not much happens in the way of climate interactions in the thin atmosphere above 11 kilometers (which is why plane flight is safer from the "weather" at these altitudes).

Figure AB1 gives a plot of temperature versus altitude. The temperature reaches a minimum around 11 km and the prior slope is linear with altitude. At much higher altitudes UV absorptions (by N_2 and O_2) produce "hot" ions that are too far apart to achieve a standard "Boltzmann equilibrium." Eighty percent of all molecules are below 12.3 km.

The "normalized" altitude Z' is defined by the fractional count of molecules, and so it goes from Z'=0 to Z'=1.0, rather than from Z=0 km to infinity km.

A one square meter area of earth has a radial expansion with altitude, but at 11 km that "expansion" is only a factor of 1.0035, and so this area correction is not included in any of the calculations. (The "cone" is represented by a "tube.")

Appendix B

Figure AB1: *Temperature K vs. Altitude Z (km): K(0) = 288.5K*

Legend:
- K vs. Z (K(0) = 288.5K)
- Extension of K vs. Z
- Power (W/m²) vs. Z

Each ⊢——⊣ spans 20 % of the molecular count

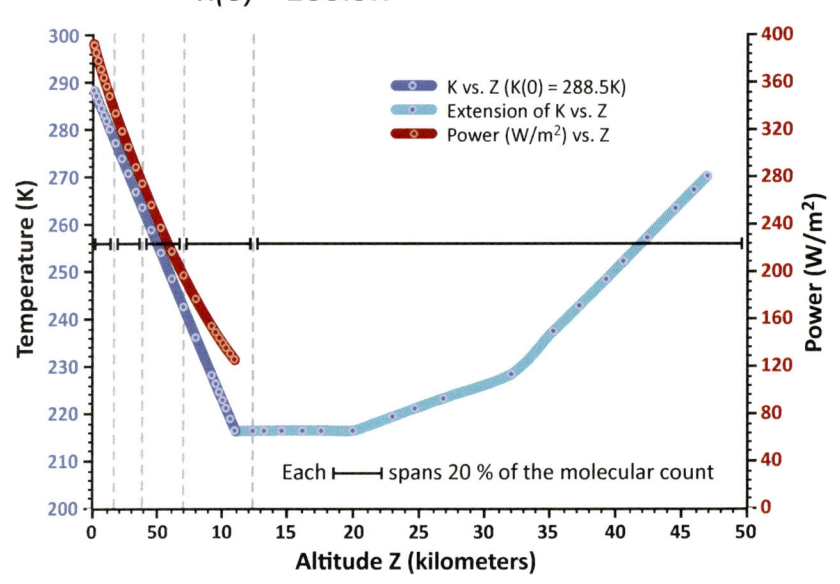

Figure AB2: *Z(km) vs. Z' (normalized)*

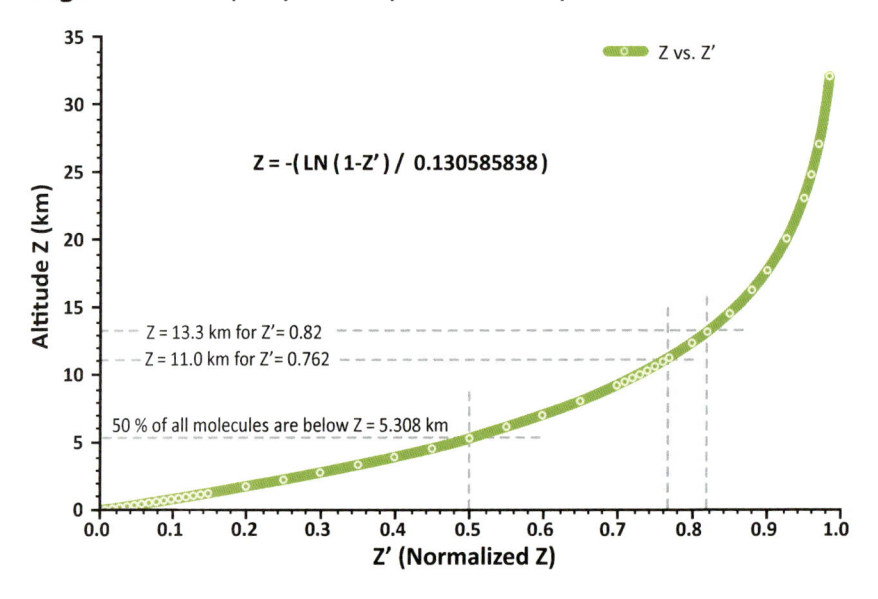

$Z = -(\text{LN}(1-Z')) / 0.130585838$

- Z = 13.3 km for Z' = 0.82
- Z = 11.0 km for Z' = 0.762
- 50 % of all molecules are below Z = 5.308 km

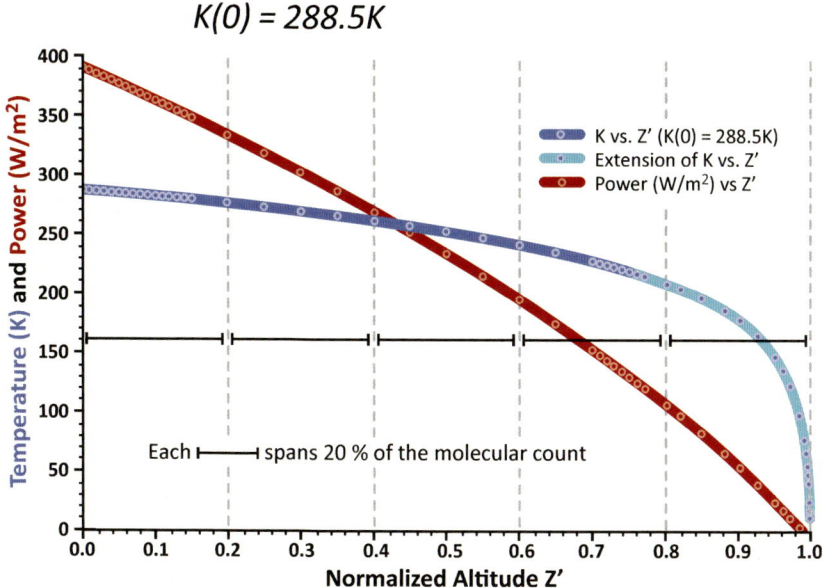

Figure AB3: Temperature K vs. Normalized Altitude Z':
K(0) = 288.5K

Legend:
- K vs. Z' (K(0) = 288.5K)
- Extension of K vs. Z'
- Power (W/m²) vs Z'

Each ⊢——⊣ spans 20 % of the molecular count

X-axis: Normalized Altitude Z'
Y-axis: Temperature (K) and Power (W/m²)

By knowing that 50% of the molecular count is below 5.308 km, and assuming an exponential fit, the Z to Z' conversion is **Z = -(LN(1-Z')/0.13058583)**. 0.13058583 is equal to -LN(0.5)/5.308.

Any other fit would be acceptable by the analysis, but this is the one employed.

P and K vs Z'

Figure AB3 shows P and K vs. Z' with K(0) = 288.0K.

Whereas K vs. Z is linear, K vs. Z' has a curvature. (This curvature is properly represented in the full calculations.)

The regions beyond the universally defined linear slope for K(0) vs. Z have the linear "slope" gradually modified so that K(Z' = 1.0)

Figure AB4: $K(Z)$ for CO_2 for increasing $K(0)$ and with $K(11.0\ km)$ fixed at 216.51K

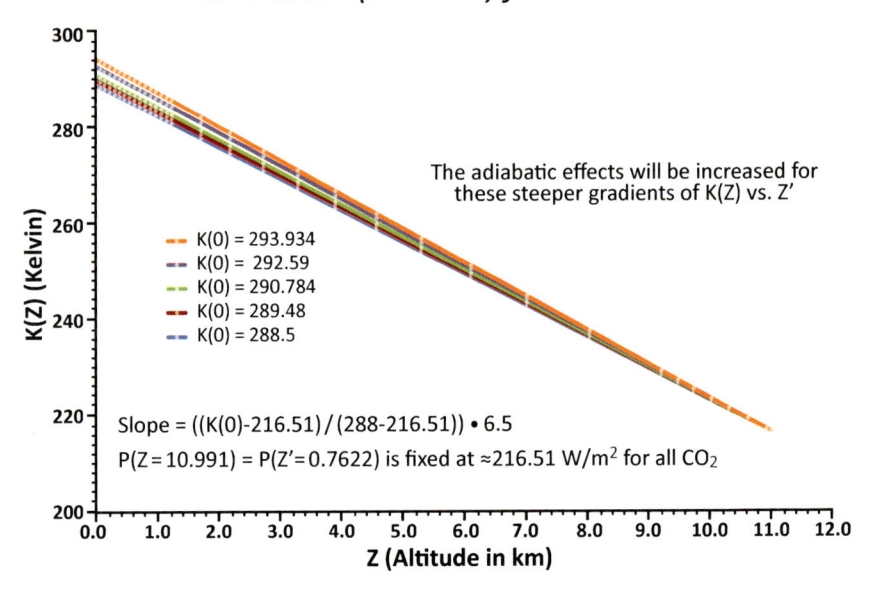

- K(0) = 293.934
- K(0) = 292.59
- K(0) = 290.784
- K(0) = 289.48
- K(0) = 288.5

The adiabatic effects will be increased for these steeper gradients of K(Z) vs. Z'

Slope = ((K(0)-216.51)/(288-216.51)) • 6.5

$P(Z = 10.991) = P(Z' = 0.7622)$ is fixed at ≈216.51 W/m² for all CO_2

Figure AB5: $K(Z')$ for CO_2 for increasing $K(0)$ and with $K(0.7622)$ fixed at 216.51K

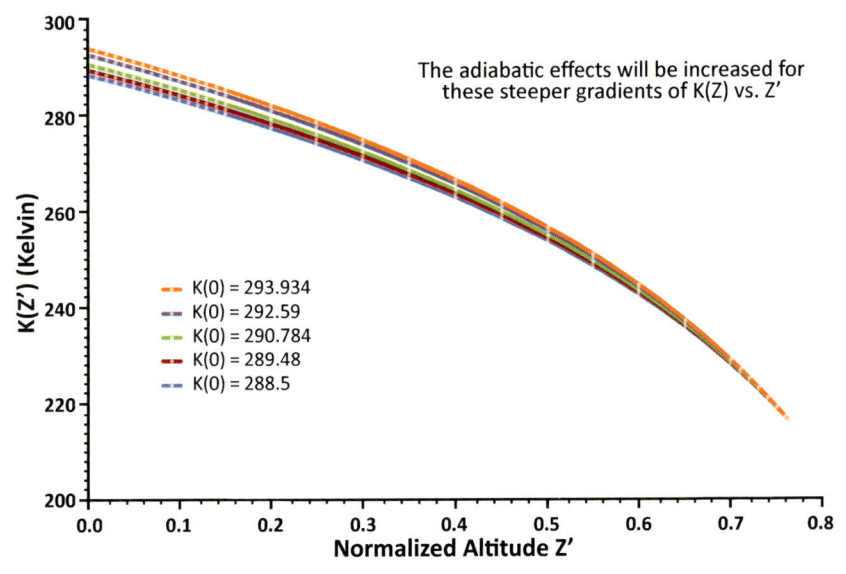

- K(0) = 293.934
- K(0) = 292.59
- K(0) = 290.784
- K(0) = 289.48
- K(0) = 288.5

The adiabatic effects will be increased for these steeper gradients of K(Z) vs. Z'

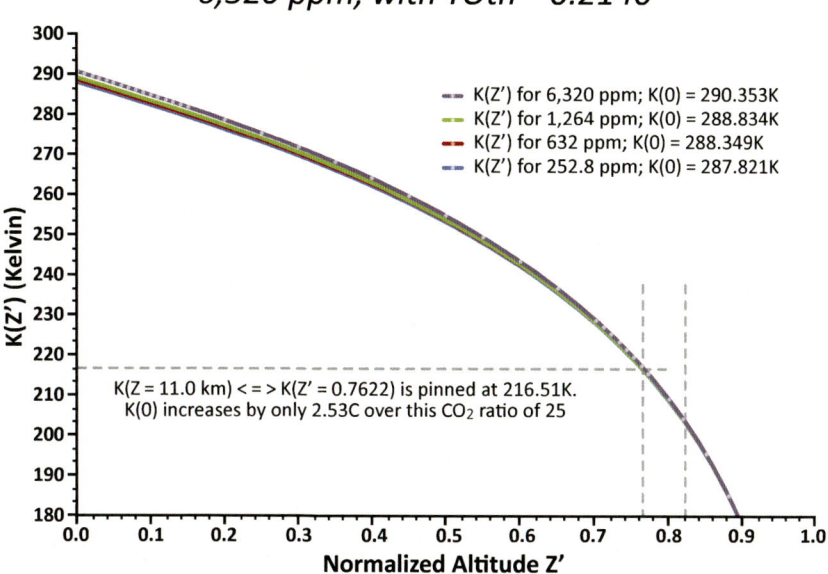

Figure AB6: $K(Z')$ for CO_2 from 252.8 ppm to 6,320 ppm, with TOth = 0.2140

Legend:
- $K(Z')$ for 6,320 ppm; $K(0)$ = 290.353K
- $K(Z')$ for 1,264 ppm; $K(0)$ = 288.834K
- $K(Z')$ for 632 ppm; $K(0)$ = 288.349K
- $K(Z')$ for 252.8 ppm; $K(0)$ = 287.821K

$K(Z = 11.0$ km$) <=> K(Z' = 0.7622)$ is pinned at 216.51K. $K(0)$ increases by only 2.53C over this CO_2 ratio of 25.

Y-axis: $K(Z')$ (Kelvin)
X-axis: Normalized Altitude Z'

equals 0.0K. These extensions are "cosmetic" extensions of the Boltzmann thermalization at lower altitudes, but the extensions have insignificant effects on the actual modeling calculations. BBR absorptions beyond $Z' = 0.82$ ($Z = 13.13$ km) are negligible.

K vs. Z demonstrates a fixed temperature of 216.5K at $Z' = 0.7622$ ($Z = 11.0$ km).

The present modeling over a wide range of $K(0)$ pins the temperature to 216.5K at $Z = 11.0$ km ($Z' = 0.7622$) for all $K(0)$. This, of course, changes the slope vs. Z, but that change in slope is accommodated into the K vs. Z' plot. Beyond $Z' = 0.7622$, $K(Z)$ has a fixed slope (and fixed $K(Z')$ values) and is independent of $K(0)$.

Figures AB4 through **AB6** show K vs. Z and K vs. Z' for a range of $K(0)$ values.

Appendix C: Organization of the spreadsheet and of the individual slices

Inputs

The first requirements are the selection of CO_2 ppm & the table of TCO_2(WNmid) for that ppm, followed by selection (or initial guess) of background TOther (which is a constant for all WN). Tables for TH_2O(WNmid) can also be inserted, but H_2O insertions are not considered here, except in so far as H_2O can be necessarily considered as a component of Other.

Guess a K(0) (or employ an entry with a known value of K(0)).

PreSelection of Z' boundaries for each slice. Typically, ΔZ' has been chosen as 0.01 (1%) or 0.05 (5%).

SubSpreadsheet calculation of K(Z') and P(Z') for each Z' edge and average valuations for K(Z'mid) and P(Z'mid). All four previously unassigned values are automatically entered for each slice.

A column of Tcomb values for each WN is obtained by multiplying TCO_2(WN) by TOth, and Tcomb(WN) is partitioned into Tcombi(WN) for each slice i. All of these values are for 300K.

Planck values [Pl(K(Z), WNmid)] • ΔWN give the full Planck power for each WN band, and the sum over all WN bands equals the full P(K(Z))

Multiplications with the temperature dependence of the Planck expression provide the temperature correction. (It may not be perfect, but it is what was used. Tcomb(WN, 300K) is already a *cumulative effect across neighboring WN absorption lines* within each band)

BBRin (i.e., BBR outputs from the previous slice, BRout(Z', WNmid) [or the full Planck(K(0), WNmid) for the first slice] are also inputs

These BBR (WN) inputs to each slice retain the WN "notching" losses from all previous slices. There are more details in **Appendix D**.

Each slice has 26 columns and about 100 rows, one for each WN plus others for data collection.

Formulas

- Full Planck value, Watts per square meter per wavenumber (W/m² per WN), for {K(0), WNmid}:

  $= \pi \cdot (1E8) \cdot (2 \cdot h \cdot c^2) \cdot (WNmid^3) / (EXP(100 \cdot h \cdot c \cdot WNmid/(k_B \cdot K(0)))-1)$ W/m²/(cm⁻¹)

  where h = Planck constant, k_B = Boltzmann constant, c = speed of light, WN = wavenumber

- Full Planck value for the input to each slice {Z', WNmid}, likewise

  $= \pi \cdot (1E8) \cdot (2 \cdot h \cdot c^2) \cdot (WNmid^3) / (EXP(100 \cdot h \cdot c \cdot WNmid/(k_B \cdot K(Z')))-1)$

- Multiplying each of these by the width, ΔWN, for each WN band, and summing, will always equal the maximum Boltzmann BBR power flux, $5.67 \cdot (K(Z')/100)^4$.

- Tcombi(WN) for each slice (prior to temperature correction) is dependent on the ΔZ' width of the slice:

  *Slice **Tcombi(WN) = EXP(-ΔZ' • LN(1/Tcomb))***

  PROD(Tcombi) = Tcomb is the full transmission (for each WN)

Global Warming Temperatures and Projections

Figure AC1: Interactions within each atmospheric slice: BBR absorption, cumulative upwards adiabatic reduction and cumulative downwards growth of the "Blanket"

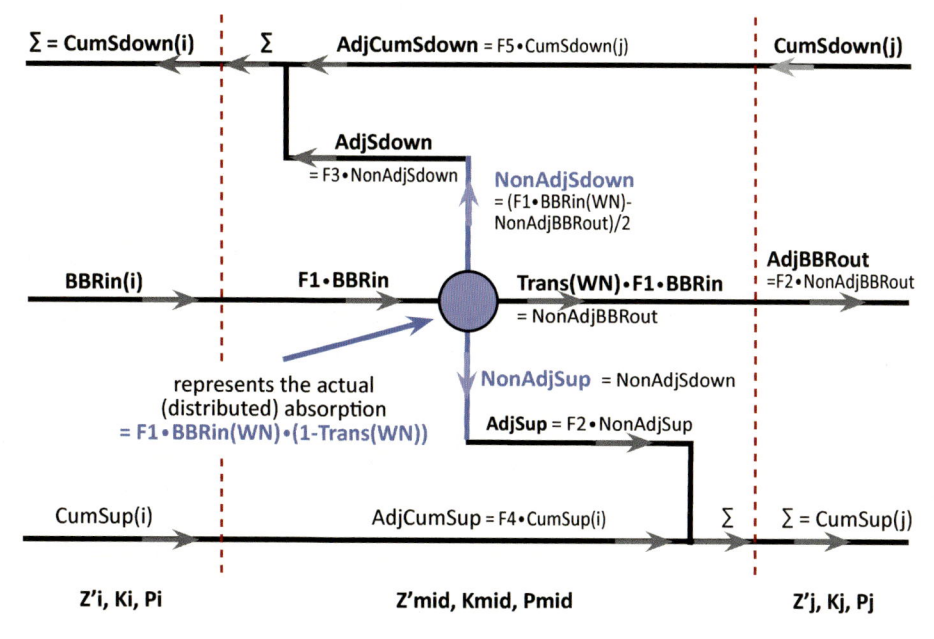

```
Σ = CumSdown(i)  ←  Σ  ←  AdjCumSdown = F5•CumSdown(j)  ←  CumSdown(j)

                          AdjSdown
                          = F3•NonAdjSdown    NonAdjSdown
                                              = (F1•BBRin(WN)-
                                              NonAdjBBRout)/2
                                                                    AdjBBRout
BBRin(i)  →  F1•BBRin  →  ⬤  Trans(WN)•F1•BBRin  →  =F2•NonAdjBBRout
                              = NonAdjBBRout

        represents the actual    ↓ NonAdjSup = NonAdjSdown
        (distributed) absorption
        = F1•BBRin(WN)•(1-Trans(WN))    AdjSup = F2•NonAdjSup

CumSup(i)  →  AdjCumSup = F4•CumSup(i)  →  Σ  →  Σ = CumSup(j)

Z'i, Ki, Pi       Z'mid, Kmid, Pmid              Z'j, Kj, Pj
```

for all molecules at 300K. For example, for 50 slices of thickness 0.02:

$$[EXP(-0.02 \cdot LN(1/Tcomb))]^{50} = Tcomb.$$

- Each slice i is assumed to have a uniform temperature represented by Kmid = (K(Z'lo)+K(Z'hi))/2, and an input Pmid = (P(Z'lo)+P(Z'hi))/2, with P consisting of both remanent BBR and thermalized photons. (Kmid and Pmid values have been pre-supplied and depend on K(0).)

- The BBR flux is processed within each WN band. {The Sum across all ΔWN equals the total BBR flux but the absorptions (notches) are differentiated from the full Planck distribution.}

- Definitions for Slice ij (between Z'i and Z'j):
There are multiple flux values tracked within each slice. **Each WN band** has calculations performed as shown in **Figure AC1**.
    * **BBRin** (equal to AdjBBRout from slice hi)
    * **NonAdjBBRout** (equal to post-adiabatic BBRin minus the absorption)
    * **AdjBBRout** (equal to BBRin for the jk slice)
        * **NonAdj** means "not adjusted" for the adiabatic effect
        Adj means "adjusted" for the adiabatic effect
    * **Sdown** and **Sup** (Sup+Sdown = 2S is the total NonAdj absorption for each band of WN)
    * **AdjSup** (lowered value of Sup at Zj because of the adiabatic effect)
    * **AdjSdown** (increased value of Sdown at Z'i because of the reverse adiab effect)
    * **CumSup** (the "cumulative" values at Z'i and Z'j of all previous AdjSup values)
    * **CumSdown** (the "cumulative" values at Z'j and Z'i of all upper AdjSdown values)

Wherever there is an "Adj" there must be a multiplying factor that provides the adiabatic adjustment.

The Fn terms are the adiabatic factors for that slice (and for each WN of the BBR flux).

In the calculation of absorption, the BBRin term is multiplied by a slice transmission factor that produces what the post-absorption value would be halfway into the slice at Kmid. This average provides an accurate 2S NonAdj absorption throughout the entire slice.

Summations are over all WN. The remaining (post-absorption and post-adiabatic) BBR continues to be tracked forward (upwards) and retains its WN bands.

- Pl(K,WNmid) is the Planck factor for a particular K(Z') and a particular WN.

- Slice 'ij' BBRin(WNmid) = 'hi' AdjBBRout(WNmid)

- AdjBBRin(Zi',WNmid) = F1 • BBRin(WN), with
    F1 = (Pl(K(Z'i,WNmid)+Pl(K(Z'j,WNmid))/
    [2 • Pl(WN,K(Z'i)]) (<1.0)

- NonAdjBBRout
    = *In-slice Transmission for each WNmid band*
    = *[(Tcombi(WNmid)) • ΔWN] • F1 • BBRin(WNmid)*

- AdjBBRout(Zj',WN)
    = F2 • NonAdjBBRout
    = F2 • [(Tcombi(WNmid)) • ΔWN] • F1 • BBRin(Wnmid)
    = 'jk' BBRin(Zj'),

- F2 = [2 • Pl(WN,K(Z'j)]/[Pl(WN,K(Z'i))+Pl(WN,K(Z'i))] (<1.0)

- **In-slice Absorption for each WNmid band**
    = [(1-Tcombi(WNmid)) • ΔWN] • F1 • BBRin(WNmid)

- The post-absorption emission is divided into two equal components, one half being a thermalized (ultimately distributed across all WN) emission upward and one half being an emission downward (NonAdjSup and NonAdjSdown).

- NonAdjSup = NonAdjSdown
    = 0.5 • (in-slice absorption for each WNmid band) = Smid

- AdjSup = F2 • NonAdjSup = F2 • Smid
    F2 is employed twice, once for the remanent post-absorption BBR and once for the upwards half of the thermalized emissions after the absorptions.

- AdjSdown = F3 • Smid, with
    F3(WNmid, slice ij) = 1/F1   (>1.0)
    and AdjSdown merges with the Pass-through blanket at Z'i.

- **F4(all WN, slice ij) = [P(K(Z'j))]/[P(K(Z'i))]**   (<1.0)
    refers to the composite sum for all WN and applies to the upwards pass-through of the thermal flux at Z'j

- **F5(all WN, Slice ij) = 1/F4**   (>1.0)
    applies to the pass-through magnification of thermalized Blanket flux from higher altitude slices

Actual parameter values for Slice Z' = 0.20 to 0.25

Figure AC2 gives actual data for a specific slice, where CO_2 = 634 ppm, K(0) = 288.349K and Other = 0.2140. The data parallels the notation from **Figure AC1. However, this data is the Sum over all WN.**

- BBRin at Z' = 0.20 = 214.5031
- Effective BBRin at Z' = 0.225 = 209.5484
- NonAdjBBRout = 190.2074
- AdjBBRout = BBRin at Z' = 0.25 = 185.7043
- NonAdjSup = NonAdjSdown = 9.6705
- Direct absorption = 19.3410
- Adiabatic Loss = 4.5031

- AdjSup at Z' = 0.25 = 9.447
- CumSup at Z' = 0.20 = 59.8903
- AdjCumSup at Z' = 0.25 = 57.1487
- CumSup at Z' = 0.25 = 66.5957

- AdjSdown at Z' = 0.20 = 9.896
- CumSdown at Z' = 0.25 = 58.0629
- RevAdjCumSdown at Z' = 0.20 = 60.8483
- CumSdown at Z' = 0.20 = 70.7747

Additional details on absorption

Figures AC1 and **AC2**, schematics of the "interactions within each slice" are remiss in their explanation of the blanket flow. The blanket development seems to imply it is the dual of Sup, but since it has

Global Warming Temperatures and Projections

Figure AC2: Calculations within a single slice, from Z' to Z' = 0.25, representing the sum over all WN (CO$_2$= 632 ppm, K(0) = 288.349K, and Other = 0.2140)

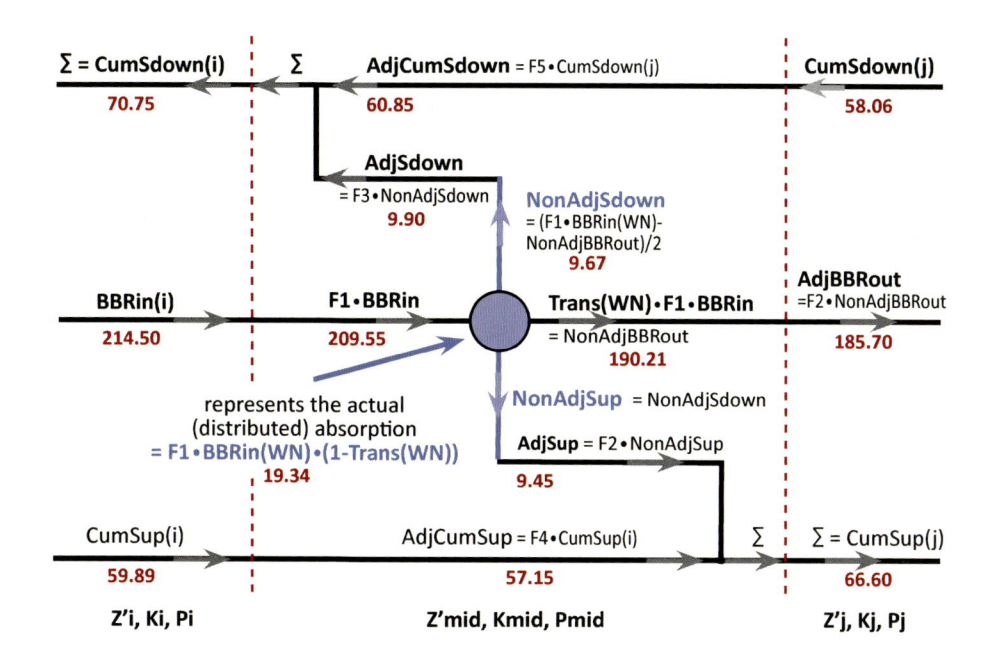

Σ = CumSdown(i)	Σ	AdjCumSdown = F5•CumSdown(j)		CumSdown(j)
70.75		60.85		58.06

AdjSdown = F3•NonAjSdown
9.90

NonAdjSdown = (F1•BBRin(WN) − NonAdjBBRout)/2
9.67

BBRin(i)		F1•BBRin	Trans(WN)•F1•BBRin	AdjBBRout =F2•NonAdjBBRout
214.50		209.55	= NonAdjBBRout 190.21	185.70

represents the actual (distributed) absorption
= F1•BBRin(WN)•(1−Trans(WN))
19.34

NonAdjSup = NonAdjSdown

AdjSup = F2•NonAdjSup
9.45

CumSup(i)		AdjCumSup = F4•CumSup(i)	Σ	Σ = CumSup(j)
59.89		57.15		66.60
Z'i, Ki, Pi		Z'mid, Kmid, Pmid		Z'j, Kj, Pj

different values than Sup, Sup is not its true dual. A proper representation of the {Sup, Sdown} pairing is shown in the following schematic. Sup and Sdown are indeed exact dual pairs representing the thermalized results of post-absorption emission.

The blanket has its own dual in much the same way as short wave length incident solar absorption produces its own isotropic emissions. **The upward BBR flux at the surface immediately becomes a matching component of the Blanket.**

Figure AC3 is an updated "refresh" of the previous figures. There are no new Fn's.

As already discussed, the schematic presents the full-fledged model for fluxes associated with a single ΔZ' slice. To the left is lower

Figure AC3: *Interactions within each atmospheric slice: BBR absorption, development of thermalized (up and down) radiation, and cumulative downwards growth of the "Blanket"*

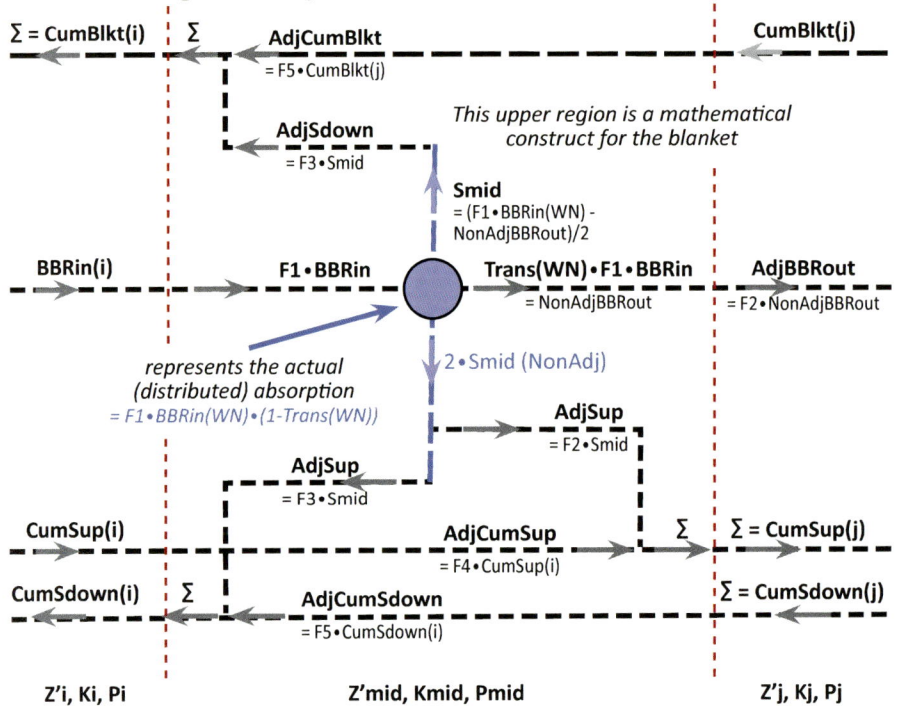

Σ = CumBlkt(i) Σ AdjCumBlkt = F5•CumBlkt(j)   CumBlkt(j)

This upper region is a mathematical construct for the blanket

AdjSdown = F3•Smid

Smid = (F1•BBRin(WN) − NonAdjBBRout)/2

BBRin(i)   F1•BBRin   Trans(WN)•F1•BBRin = NonAdjBBRout   AdjBBRout = F2•NonAdjBBRout

represents the actual (distributed) absorption = F1•BBRin(WN)•(1−Trans(WN))

2•Smid (NonAdj)

AdjSup = F2•Smid

AdjSup = F3•Smid

CumSup(i)   AdjCumSup = F4•CumSup(i) Σ Σ = CumSup(j)

CumSdown(i) Σ AdjCumSdown = F5•CumSdown(i) Σ = CumSdown(j)

Z'i, Ki, Pi   Z'mid, Kmid, Pmid   Z'j, Kj, Pj

Z'; to the right is higher Z'. As the BBR flux progresses upward some is lost to adiabatic expansion and some is absorbed. **However, no energy is lost by an absorption; the only aspect that may be lost is its directionality.** *Energy loss* occurs only as thermodynamically defined adiabatic loss or as true "escape".

The upper half of the schematic concentrates on the mathematical construct of the Blanket development. There is no actual "flow." What was a single track at the bottom is now a double track representing the total thermalization. Their combination is 2 • AdjCumS, with equal upwards and downwards radiation.

The down path, AdjCumBlkt, is difficult to calculate directly since it starts far off to the right at Z'cutoff. AdjCumBlkt is somewhat less than AdjCumSup—which begins its calculations at Z' = 0. At Z = Z'cutoff, Blkt = 0, but AdjCumSdown(Z'cutoff) equals 2 • AdjCumS(Z'cutoff)/2.

Figure 9 in the main text should now have more meaning. The red tent is the developing Blanket, and the same area beneath is a portion (39%) of the BBR.

BBR absorption

In **Figure AC3**, in which the parameters of a slice are simplified as {Z'mid, ΔZ', Kmid, Pmid}, it is also assumed that the "average" BBR (WN) that is being absorbed is represented by what its adiabatic value would be in the center of the slice. That is, the available BBR is actually greater at the incoming edge and less at the outgoing edge, and the average to apply to the entire slice is the adiabatically determined value midway between the edges.

The actual post-absorption emission is isotropic, and so before the "up" and "down" components experience any adiabatic effects as thermalized radiation their values, nonAdjSup and nonAdjSdown, are equal. Subsequently, each experiences opposing adiabatic effects before crossing the edges and co-joining their mainstream thermalized flows.

The actual absorption (WN), without any consideration of prior or subsequent adiabatic effects, equals:

$$nonAdjSup(WN) + nonAdjSdown(WN) = 2 \cdot Smid(WN)$$

K(0) is adjusted until CumBlkt(0) equals the required Blkt(0) for that K(0). It is this "Do Loop" that guarantees the equilibrium of the solution.

The thought is simple; the bookkeeping is tricky. The chosen granularity for a slice is sufficient to maintain valid boundary conditions since no disturbing offsets appear in the full plots. When a slice was further sliced into five slices, the final outputs were the same within three significant figures.

Table AC1

CO_2 (ppm)	K(0) (Kelvin)	P(0) (W/m²)	CumAbs (Z'=0.82)	Blkt (0) (W/m²)	Blkt(0) /CumAbs
0	284.722	372.621	199.101	133.62	0.671
632	288.349	391.974	236.832	152.976	0.646
1,264	288.834	394.618	241.933	155.619	0.643
6,320	290.353	402.985	258.434	163.986	0.635

The cumulated sum of BBR absorptions across all WN bands for different levels of CO_2. In all cases, TOth = 0.2140.

Figure AC4 shows the cumulated sum (summing across WN bands, plus summing from slice to slice) of the actual absorptions in each slice. **Although the remanent BBR (WN) entering each slice has experienced pre-absorption adiabatic loss,** this cumulated sum represents the increments of BBR absorption at the points of absorption, since any subsequent adiabatic effects do not affect the simple sum of the post-absorption thermalized emissions. {The remanent BBR at Z'= 0.82 is only 1.0 to 2.0 W/m². Subsequent post-absorption downward-directed emissions experience no adiabatic gain until they are below Z'= 0.82.}

For Z'<0.82 the remanent BBR(WN) after absorption experiences adiabatic loss from {Z'mid, Plmid} to {Z'up, Plup}. This BBR(WN) input to the next slice also experiences an adiabatic loss from {Z'down, Pldown} to {Z'mid, Plmid} of the *next* slice, and this identifies the average BBR(WN) available for absorption in *that* slice.

More discussion will follow, but a summary table, with TOth = 0.2140 for all CO_2 is presented in **Table AC1**.

Global Warming Temperatures and Projections

Figure AC4: *Cum BBR Direct Absorption (w/o adiab);*
(CO_2 = 632, 1,264, and 6,320 ppm, TOth = 0.2140)

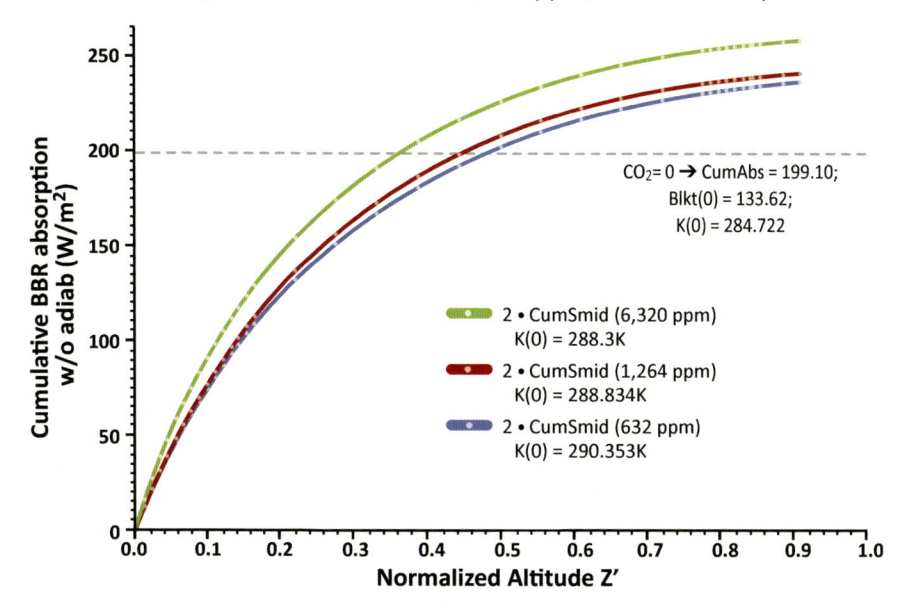

CO_2 = 0 → CumAbs = 199.10;
Blkt(0) = 133.62;
K(0) = 284.722

2 • CumSmid (6,320 ppm) K(0) = 288.3K
2 • CumSmid (1,264 ppm) K(0) = 288.834K
2 • CumSmid (632 ppm) K(0) = 290.353K

Y-axis: Cumulative BBR absorption w/o adiab (W/m²)
X-axis: Normalized Altitude Z'

	6.320 ppm; K(0) = 290.353K TOth = 0.2140	1,264 ppm; K(0) = 288.834K TOth = 0.2140	632 ppm; K(0) = 288.349K; TOth = 0.2140
Cumulative BBR Absorption = Sum of 2 • Smid for each slice	258.43	241.93	236.83
P(0)	402.98	394.62	391.97
P(0.82)	97.23	97.23	97.23
BBR(0.82)	17.72	19.41	20.00
AdjCumSup(0.82)	39.55	38.32	37.93
[P(0)-P(0.82)]-BBR(0.82)-AdjCumSup(0.82)	248.48	239.66	236.81
Blkt(0)	163.986	155.62	152.98
Blkt(0)/CumBBR Abs	0.635	0.643	0.646

[(P(0, 1,264)-P(0, 632)]/[(CumAbs(0.82, 1,264)-CumAbs(0.82, 632)] = 0.52;
[(Abs(0.82, 1,264)-Abs(0.82, 632)] /[(K(0, 1,264)-K(0, 632)]= 5.45 W/m² /K

[(P(0, 6,320)-P(0, 1,264)]/[(CumAbs(0.82, 6,320)-CumAbs(0.82, 1,264)] = 0.51;
[(Abs(0.82, 6,320)-Abs(0, 1,264)]/[(K(0, 1,264)-K(0, 632)]= 5.51 W/m² /K

Cumulative sum of BBR absorptions

Figure AC4 shows plots of the accumulated sum of BBR absorptions with altitude. The plot is necessarily sublinear since there is a lower remanent BBR input to each slice.

These are the absorptions that create both the warming downward "blanket," via Sdown, and the upward emissions which can only be eliminated by the adiabatic effect. Nature accommodates by automatically raising the temperature gradients so that the cooling adiabatic loss rate is increased.

The ratios of CO_2 for subsequent plots are 2.0 and 5.0, for a total ratio of 10. These ratios are much greater than the much lower ratios that lately are of great concern in the media, but there is nothing disconcerting about the results. **A ten-fold increase in CO_2 is NOT increasing total absorption by 1,000 percent, but only by 9 percent.** A reasonable person would not be alarmed.

The summaries that accompany **Figure AC4** are also interesting. **Each 1.0 increase in K(0), or about 5.5 W/m² increase in P(0), requires an increase in absorption of about 11 W/m². This is nature's safety factor of two against any possible runaway,** something that is very desirable for bridges, but is not a problem here.

The ratio of Blanket at Z' = 0 to CumAbs at Z' = 0.82 (the chosen "cutoff" for adiabatic effects) consistently shows values of about 0.64. All is calm.

"Notches" in WN in the BBR distribution

The 26-column spreadsheet sections for each slice have BBR inputs and outputs for each WN band (the inputs being carry-forwards from the previous slice).

For any given K(0), and any altitude Z', the BBR values can be selected for each WN band. For the same K(0) spreadsheet, sets of BBR values can be obtained for a variety of altitudes.

Full Planck distributions are also available for each Z', each of which will have its own K(Z').

Each BBR distribution can be nested under its own Pl(K(Z'))

Global Warming Temperatures and Projections　　　**187**

Figure AC5: BBR Transmission factor vs. WN, including adiabatic loss, from Z' = 0 to slice output
{BBR T averaged over WN ±25 cm^{-1}; CO_2 = 500 atm-cm (632 ppm); TOth = 0.2140}

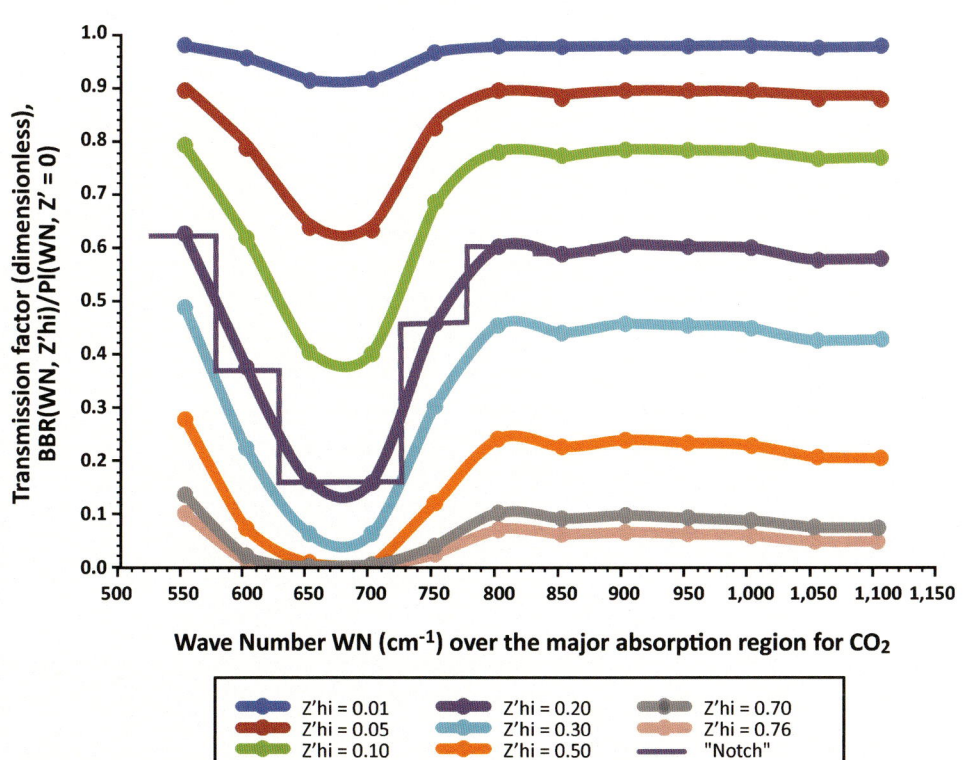

Transmission factor (dimensionless), BBR(WN, Z'hi)/Pl(WN, Z' = 0) vs. Wave Number WN (cm^{-1}) over the major absorption region for CO_2

Legend: Z'hi = 0.01, Z'hi = 0.05, Z'hi = 0.10, Z'hi = 0.20, Z'hi = 0.30, Z'hi = 0.50, Z'hi = 0.70, Z'hi = 0.76, "Notch"

distribution. The BBR value at any WN will never exceed the full Planck value. BBR notches begin to appear even at the lowest altitudes (**Figure AC5**).

It is not that there is an absence of photons having energies within any notch; it's just that these photons are "re-fill" photons from the thermalized distribution. They too are actively absorbing and re-emitting, but they do not affect the Blanket or K(0).

Without any explanation more complete than the one above, **Figure AC5** is an example of notch development as the altitude increases. (For more, see **Appendix D**.)

Adding more CO_2 will (weakly) affect absorption elsewhere but not at all in the fully developed notch. {H_2O has its own unique notches and the combination of CO_2 and H_2O produce integrated notches (not sums).}

Emphases

It is critically important, in so far as accuracy is concerned, to process the BBR WN band by WN band.

If the remanent BBR were to be treated as just a scaled down version ((P BBR)/(P overall)) of the full Planck distribution, then the most sensitive WN bands will keep indicating absorption, an absorption that is not taking place.

Absorption and emission are, of course, always taking place within the thermalized distribution, with isotropic absorption and emission, but this is of no consequence for the BBR spectrum where the input to the absorption is directional and the emission is isotropic.

Failure to analyze the absorption band by band will result in higher (false) predictions of $\Delta K(0)$ per doubling of CO_2.

The ΔWN bands are 50 cm^{-1}, and the Planck value changes only slightly over such a narrow band. Employing Pl(Wnmid) • 50 is a very good approximation to the integrated area.

Slice widths of $\Delta Z' = 0.01$ are sufficiently small. Recalculations with a sequence of five slices with $\Delta Z' = 0.002$ give results that agree perfectly within three significant digits. This also confirms the rigor of the boundary conditions.

Sums over all WN give the total BBR input and output flux values.

Appendix D: Absorption 'notches'

"Notches" of absorption as a function of WN and altitude as CO_2 is increased

Dramatic representations of the effects of CO_2 are shown in the following three triplets of figures (**AD1 through AD9**). Each triplet is for a different CO_2: 500 atm-cm (632 ppm); 1,000 atm-cm (1,264 ppm); and 5,000 atm-cm (6,320 ppm).

The WN range is 525 cm^{-1} to 1,125 cm^{-1}, with each band having a width of ± 25 cm^{-1} on either side of WNmid. The altitude selections are: Z'mid = 0.005; Z' = 0.045; Z'= 0.095; Z' = 0.195, Z'= 0.295; Z' = 0.495; Z' = 0.695; Z' = 0.755. The selected altitude height is the fixed parameter for each plot.

The first of each triplet shows the remaining transmission (at altitude Z') from the original BBR radiation (at Z' = 0) for each of CO_2's most sensitive WN bands, with additional absorptions from TOth = 0.2140. The second figure of each triplet is the dimensionless BBR transmission coefficient from Z'=0 to a slice output Z'out. T=(BBR(Z'out,WN))/(BBR(0),WN)). The third figure of each triplet

Figure AD1: BBR transmitted power per WN band relative to the input Planck value at Z' = 0, including both absorption and pre-absorption adiabatic dissipation — at increasing Z' altitudes, with CO_2 = 632 ppm and TOth = 0.2140.

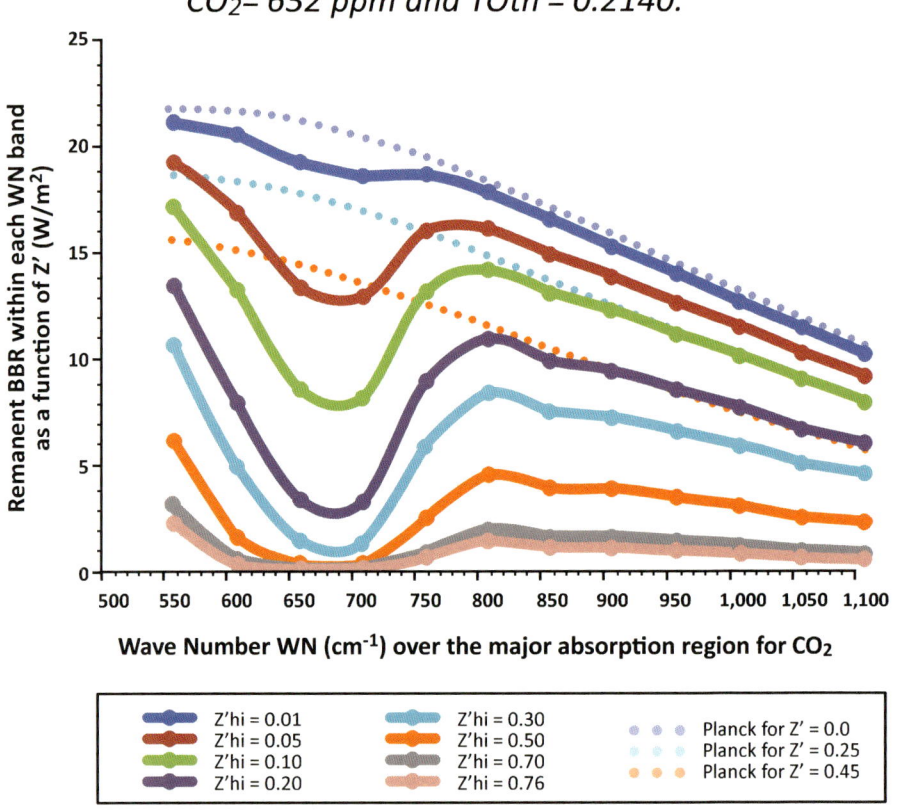

substitutes (Pl(Z'in,WN) for (BBR(0,WN). The differential transmission between slices can be observed from the third figure.

Actual Planck distributions are shown for chosen altitudes. Note that the Planck distribution provides a coverlet over each of the remanent BBR distributions. The volume between the BBR and the Planck distributions is not empty; it is occupied by isotropic

Figure AD2: BBR Transmission factor per WN band ((BBR(Z',WN)/BBR(0,WN)) vs. WN — at increasing Z' altitudes, with CO_2 = 632 ppm and TOth = 0.2140.

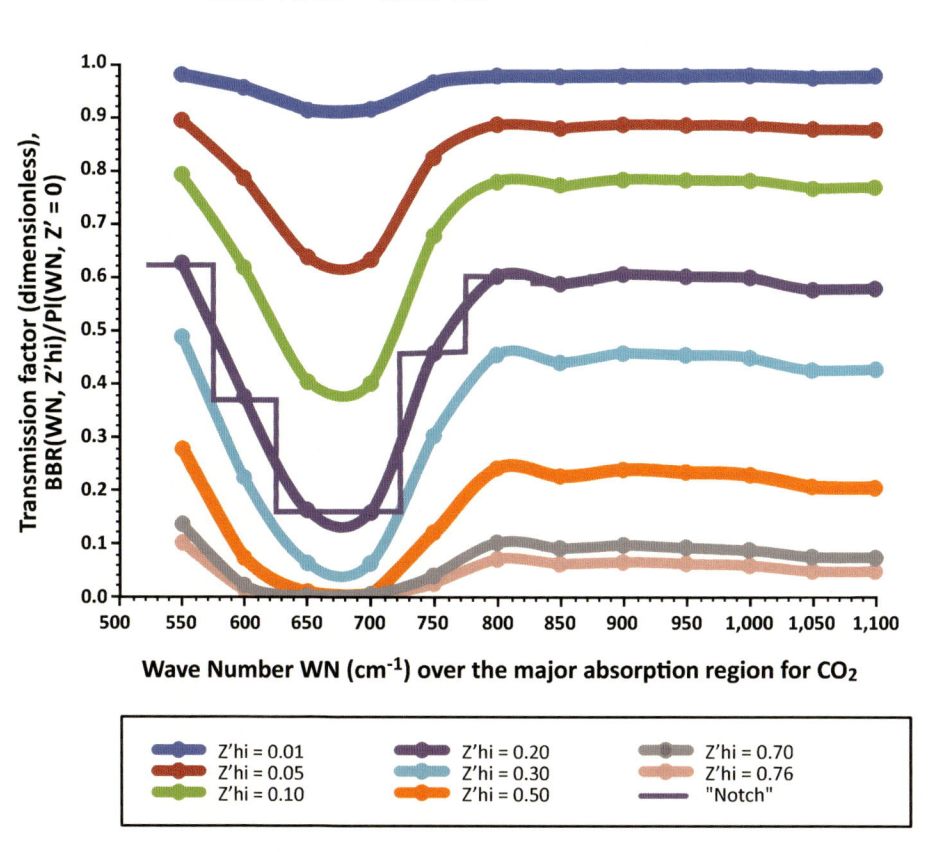

thermalized photons.

T (dimensionless) = BBRout(WNmid, Z'up, K)/ (Planck(WNmid, Z' = 0 or Z'down, K))

What is immediately obvious is the appearance of notches of reduced transmission, i.e., strong absorption. The most sensitive

Figure AD3: BBR transmission factor per WN band, including adiabatic loss, for slice output relative to the full slice input ((BBR(Z'out,WN)/(PI(Z'in,WN)) vs. WN — at increasing Z' altitudes with CO_2 = 632 ppm and TOth = 0.2140

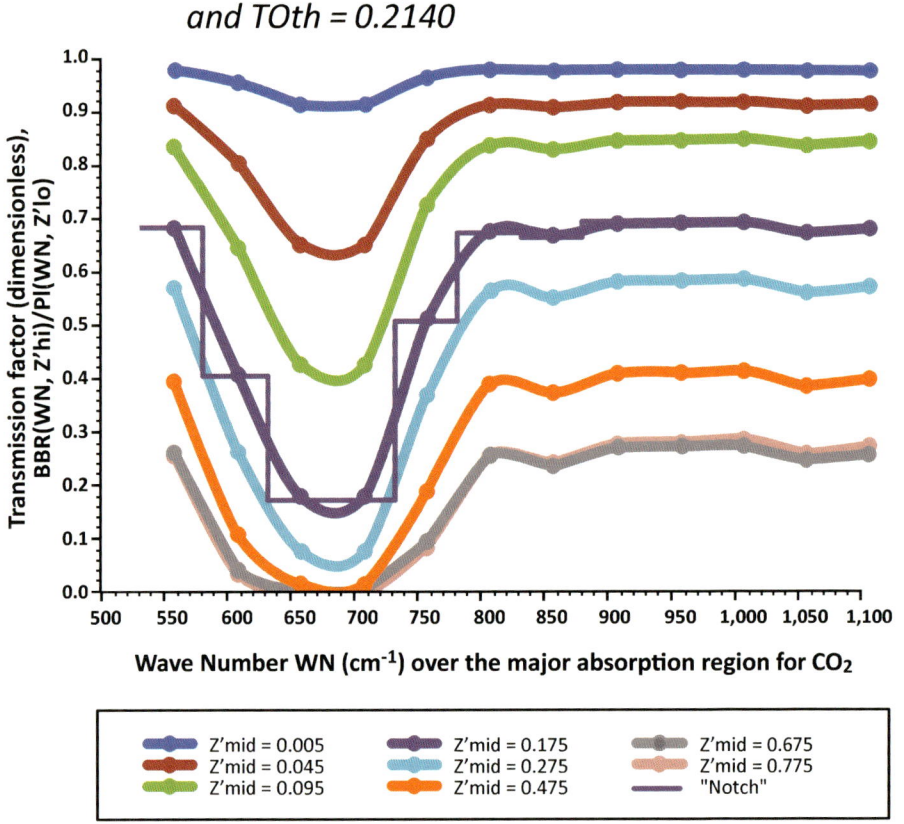

Z'mid = 0.005	Z'mid = 0.175	Z'mid = 0.675
Z'mid = 0.045	Z'mid = 0.275	Z'mid = 0.775
Z'mid = 0.095	Z'mid = 0.475	"Notch"

absorption lines are nearly "saturated" well below 632 ppm.

The notch for 632 ppm maintains only 16% transmission at a height of Z' = 0.20 (which is sweeping through an equivalent of only 126 ppm). **"Saturation" across a width of 100 cm^{-1} (625 to 725 cm^{-1}) is fully apparent at Z' = 0.475.**

The saturation notch for 1,264 ppm at Z' = 0.475 is wider (the

Global Warming Temperatures and Projections

notch is more accurately described as a trapezoid). It also maintains only 16% transmission at a height of Z' = 0.20.

The notch for 6,320 ppm **(ten times greater than 632 ppm)** has noticeably lower slopes along its sides. It also maintains only 16% transmission at a height of 0.20.

The notches are obviously becoming wider. But the depths associated with the most sensitive range of WN (650 to 700 cm^{-1}) are not changing significantly even as ppm rise. The reason is that the transmission tables do not have transmission entries below 0.001 since there can realistically never be a 100% absorption. The transmission tables for TCO$_2$ between 625 and 675 and between 675 and 725 are *catalogued as 0.001 for all CO$_2$ levels above 250 ppm.* So even though CO$_2$ ppm are higher (more absorption) the indicated (dimensionless) transmission results are not changed even as CO$_2$ levels reach 1,264 ppm. *Nevertheless, the true notches at all lower heights are deeper as CO$_2$ levels increase.*

- **Figures AD1. AD2** and **AD3:**
  CO$_2$ = 632 ppm (500 atm-cm) with Z' as a parameter

  BBR(Z',WN) (W/m^2), with Planck(K(Z'),WN) at selected Z' (W/m^2/WN)
  Transmissivity BBR(Z',WN)/BBR(0,WN)
  Transmissivity BBR(Z'out,WN)/Pl((Z'in, WN)

- **Figures AD4. AD5** and **AD6:**
  CO$_2$ = 1,264 ppm (1,000 atm-cm) with Z' as a parameter

  BBR(Z',WN) (W/m^2), with Planck(K(Z'),WN) at selected Z' (W/m^2/WN)
  Transmissivity BBR(Z',WN)/BBR(0,WN)
  Transmissivity BBR(Z'out,WN)/Pl((Z'in, WN)

- **Figures AD7. AD8** and **AD9:**
  CO$_2$ = 6,320 ppm (5,000 atm-cm) with Z' as a parameter

  BBR(Z',WN) (W/m^2), with Planck(K(Z'),WN) at selected Z' (W/m^2/WN)
  Transmissivity BBR(Z',WN)/BBR(0,WN)
  Transmissivity BBR(Z'out,WN)/Pl((Z'in, WN)

Figure AD4: *BBR transmitted power per WN band relative to the input Planck value at Z' = 0, including both absorption and pre-absorption adiabatic dissipation — at increasing Z' altitudes, with CO_2 = 1,264 ppm and TOth = 0.2140.*

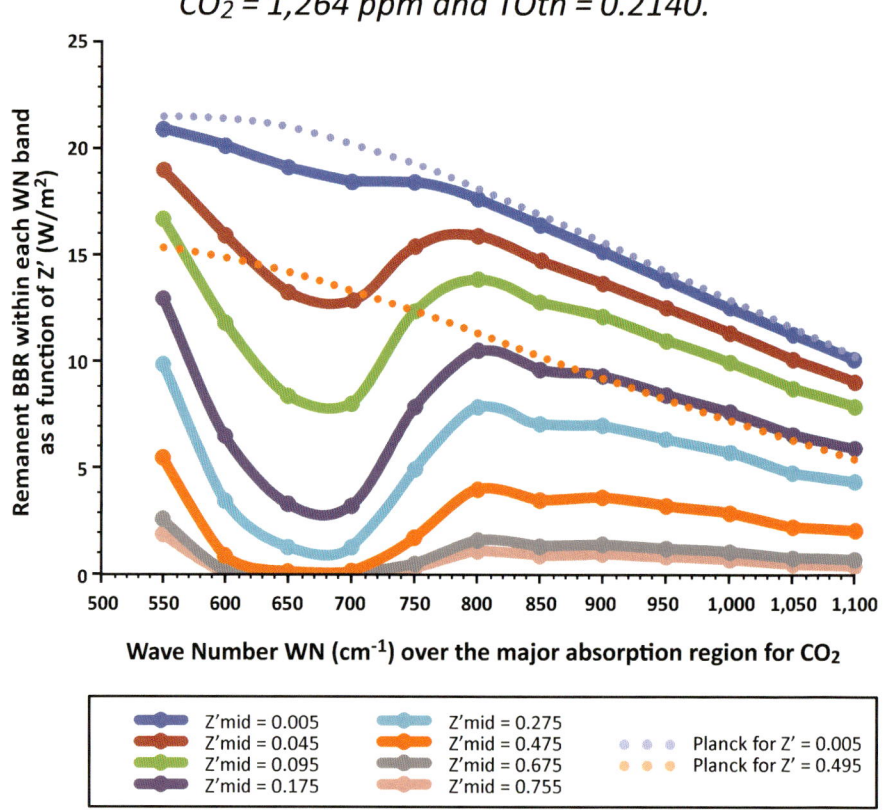

Interpreting the notch

The WN range (525 to 1,125 cm^{-1}) shown for the notches emphasizes the range within which CO_2 absorption is significant. For larger WN there are other absorbing ranges, but within those ranges there is very little power within the Planck distributions associated with

Figure AD5: BBR Transmission factor per WN band ((BBR(Z',WN)/BBR(0,WN)) vs. WN — at increasing Z' altitudes, with CO_2 = 1,264 ppm and TOth = 0.2140.

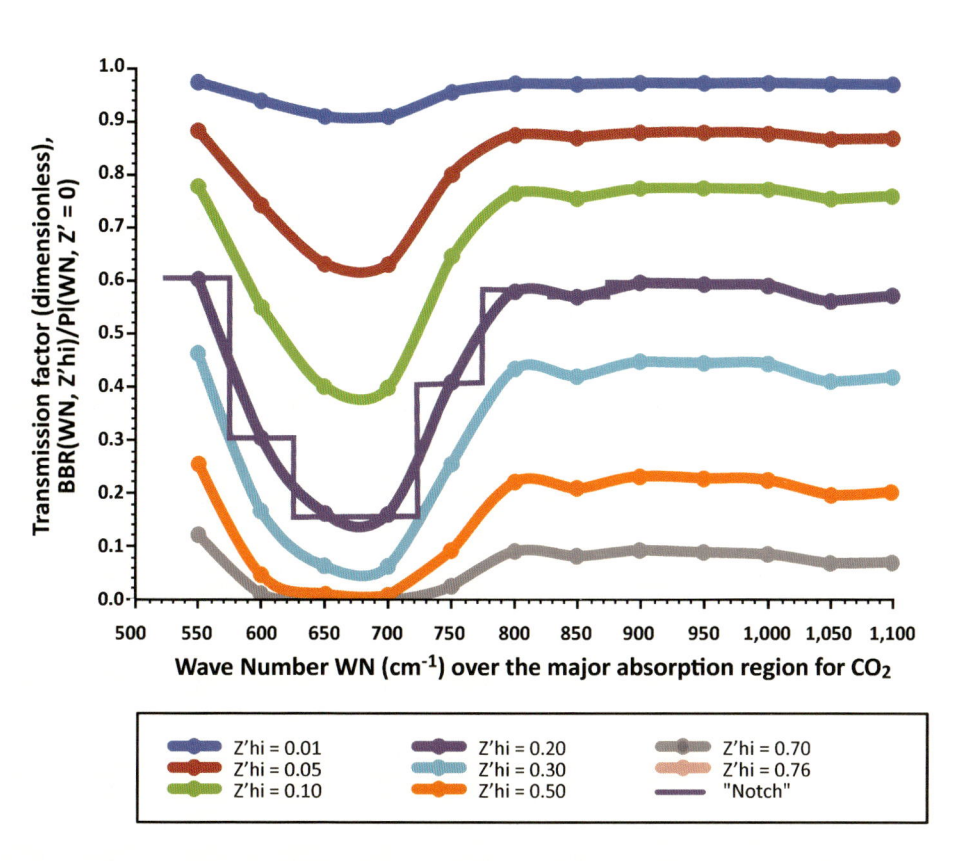

the BBR emissions.

Nevertheless, these ranges are also included within the spreadsheet calculations.

The notches shown represent altitudes from Z' = 0.01 (Z = km) to Z' = 0.76 (Z = km). The major CO_2 interactions are for WN between 575 and 775 cm^{-1}.

Figure AD6: BBR transmission factor per WN band, including adiabatic loss, for slice output relative to the full slice input ((BBR(Z'out,WN)/(Pl(Z'in,WN)) vs. WN — at increasing Z' altitudes with CO_2 = 1,264 ppm and TOth = 0.214

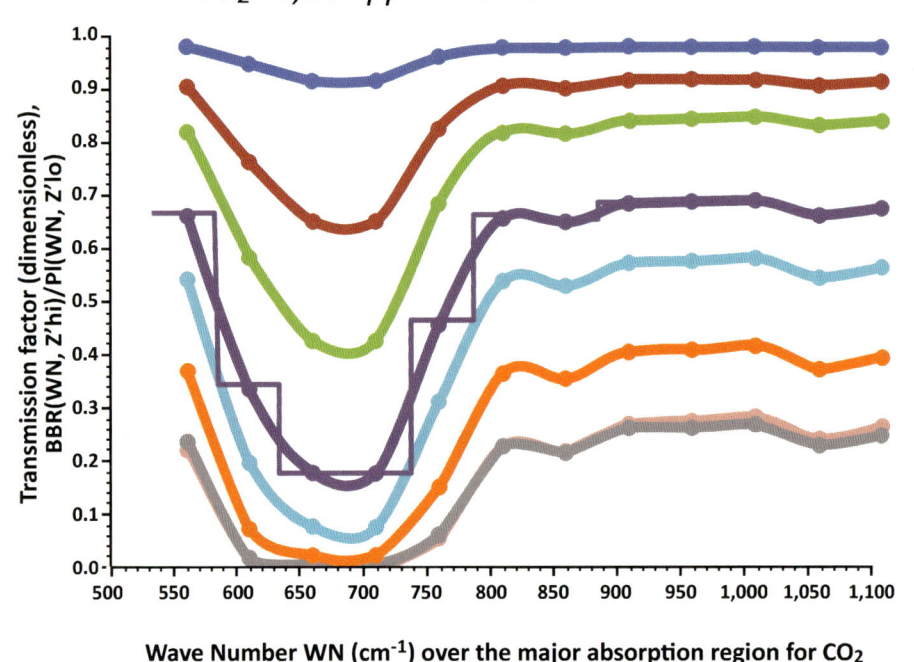

Wave Number WN (cm^{-1}) over the major absorption region for CO_2

- Z'mid = 0.005
- Z'mid = 0.045
- Z'mid = 0.095
- Z'mid = 0.175
- Z'mid = 0.275
- Z'mid = 0.475
- Z'mid = 0.675
- Z'mid = 0.775
- "Notch"

The ripples for WN values greater than 775 are associated with CO_2 but most of the loss in transmission is associated with background (non-CO_2) absorptions.

At the earth's surface, the Planck distribution power available to CO_2 is about 51.5% of the total power in the distribution. Therefore, CO_2 can never influence more than one half of the BBR emission.

Figure AD7: BBR transmitted power per WN band relative to the input Planck value at Z' = 0, including both absorption and pre-absorption adiabatic dissipation — at increasing Z' altitudes, with CO_2 = 6,320 ppm and TOth = 0.2140.

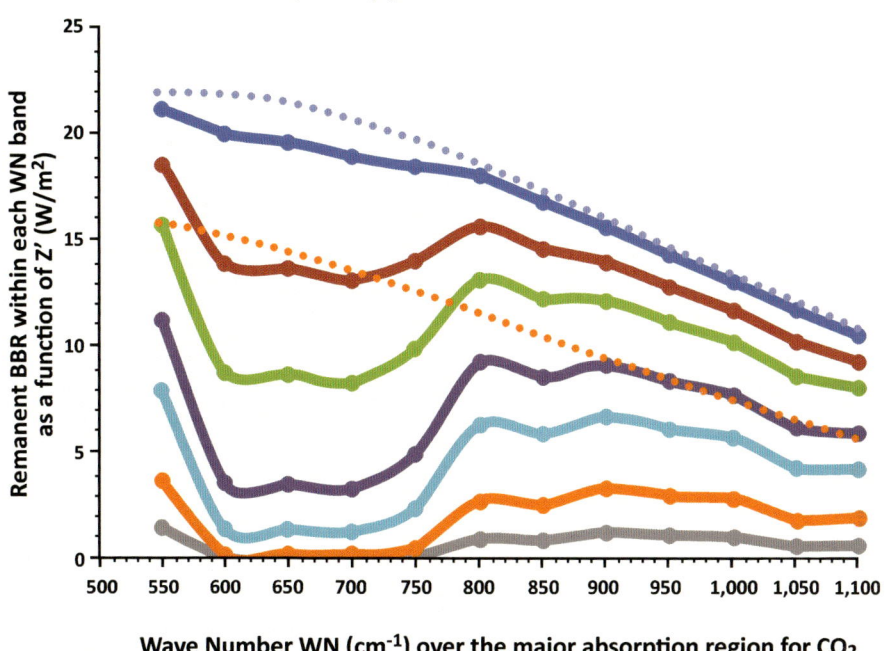

Z'mid = 0.005	Z'mid = 0.275	Planck for Z' = 0.005
Z'mid = 0.045	Z'mid = 0.475	Planck for Z' = 0.495
Z'mid = 0.095	Z'mid = 0.675	
Z'mid = 0.175	Z'mid = 0.755	

Y-axis: Remanent BBR within each WN band as a function of Z' (W/m²)

X-axis: Wave Number WN (cm⁻¹) over the major absorption region for CO_2

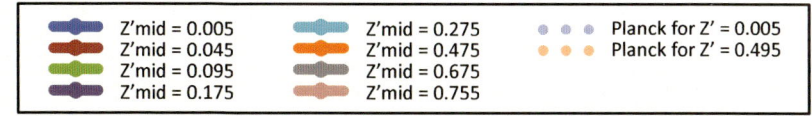

At the higher altitudes (Z = 11 km) the lower temperature and density reduce the Planck power by a factor of 3. The CO_2 interaction zone is further reduced to 45% of the total infrared power.

CO_2 absorbs in the same manner as any other molecule, but it has some particularly sensitive absorption lines within the span

Figure AD8: BBR Transmission factor per WN band (($BBR(Z',WN)/BBR(0,WN)$)) vs. WN — at increasing Z' altitudes, with CO_2= 6,320 ppm and TOth = 0.2140.

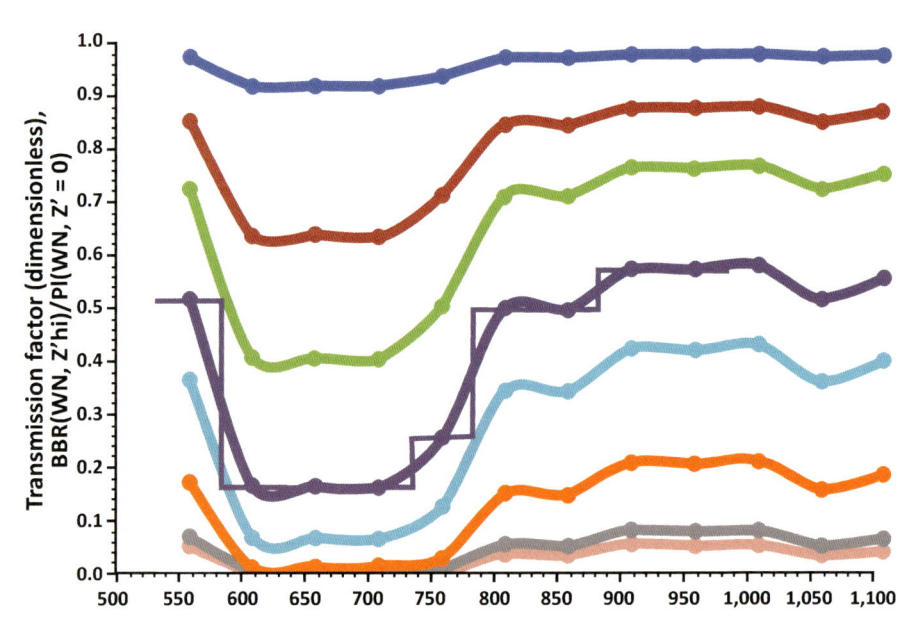

between 625 and 725 cm-1. This span represents only 10% of the Planck distribution.

Some lines — with bandwidths narrower than 50 cm^{-1} — will show "saturation" absorption (absorption factors greater than 0.99) at only 50 atm-cm (**60 ppm**) and these plots show how it takes *much less than the full height of the atmosphere* to absorb all

Figure AD9: *BBR transmission factor per WN band, including adiabatic loss, for slice output relative to the full slice input ((BBR(Z'out,WN)/(PI(Z'in,WN)) vs. WN — at increasing Z' altitudes with CO_2= 6,320 ppm and TOth = 0.2140*

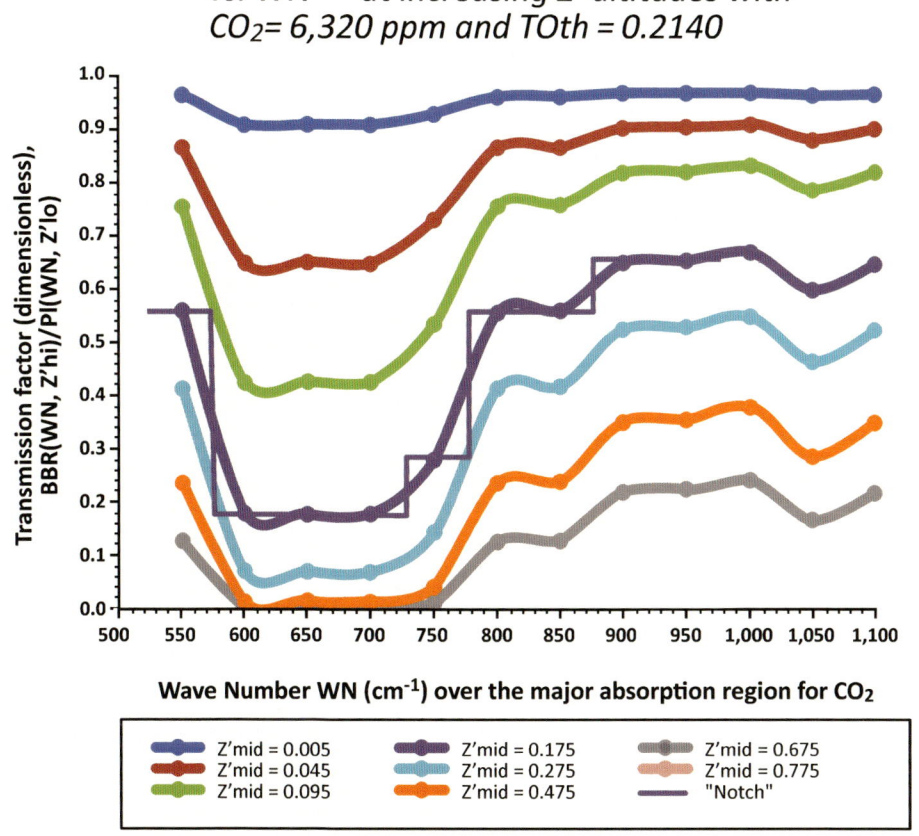

Transmission factor (dimensionless), BBR(WN, Z'hi)/PI(WN, Z'lo)

Wave Number WN (cm^{-1}) over the major absorption region for CO_2

- Z'mid = 0.005
- Z'mid = 0.045
- Z'mid = 0.095
- Z'mid = 0.175
- Z'mid = 0.275
- Z'mid = 0.475
- Z'mid = 0.675
- Z'mid = 0.775
- "Notch"

available infrared radiation *over the entire span from 625 to 725 cm^{-1} even for CO_2 levels of 100 atm-cm.*

Adding more CO_2 can do nothing, *within this span*, to enhance the absorption of black body radiation from the earth and will do nothing to enhance the blanket radiation.

Higher CO_2 levels simply allow more absorption to take place within the thousands of less sensitive absorption lines on either side of the dramatically sensitive region. Within this region CO_2 carries out its job of BBR absorption and calls it a day in so far as any additional BBR absorption.

There are no "runaway" effects, and all such statements should not be in the literature. (Nature always seeks a low energy minimum even if the path there may be dramatic — e.g., a hurricane or a even a lightning strike. The immediate days after a storm are always peaceful.)

Appendix E:
Final sets of data and their interpretation

This appendix presents results of a series of spreadsheet calculations. The first set enlarges the discussion in Section One; in this set of calculations the only variable of absorption is CO_2. Water vapor absorption, evaporation and condensation, additional blanket inputs from the Tropics, and absorptions by other molecules are captured within a TOther "background." This approach was discussed in the main text, and what is presented here are sequences of results, for the baseline requirement of (CO_2 = 380 ppm, K(0) = 288.0K, TOther = 0.2140).

Set#1: CO_2 only (a constant H_2O and convection included within TOth)

K(0) (K)	CO_2 (ppm)	TOth	Req Blkt (W/m²)	Req-Calc (W/m²)
287.821	252.8	0.21401	150.111	-0.0017
288.0	380	0.21401	151.079	0
288.350	632	0.21401	152.979	0.0010

ΔK(0) = 0.529K → 0.40K per doubling
ΔK(0) per 100 ppm = 0.14

Appendix E

K(0) (K)	CO_2 (ppm)	TOth	Req Blkt (W/m²)	Req-Calc (W/m²)
288.834	1,264	0.21401	155.617	-0.0014

$\Delta K(0) = 0.484 \rightarrow 0.48K$ per doubling
$\Delta K(0)$ per 100 ppm = 0.077

$\Delta Abs = 2 \cdot \Delta Blkt(0) = 5.28$ W/m²

K(0) (K)	CO_2 (ppm)	TOth	Req Blkt (W/m²)	Req-Calc (W/m²)
289.444	2,528	0.21401	158.797	-0.0002

$\Delta K(0) = 0.61 \rightarrow 0.61K$ per doubling
$\Delta K(0)$ per 100 ppm = 0.048

K(0) (K)	CO_2 (ppm)	TOth	Req Blkt (W/m²)	Req-Calc (W/m²)
290.353	6,320	0.21401	163.989	-0.0011

$\Delta K(0) = 0.909 \rightarrow 0.69K$ per dblng
$\Delta K(0)$ per 100 ppm = 0.0234

$\Delta K(0)$ increases from 0.40K per doubling to 0.69K per doubling. Overall, $\Delta K(0)$ has increased by 2.5K for a 25X increase in CO_2. CO_2 is clearly not a villain. The average $\Delta K(0)$ per 100 ppm is low and decreases steadily as CO_2 increases. $\Delta Abs/\Delta K(0)$ is always $\cong 11$ W/m².

The percentage increase due to CO_2 for a doubling of CO_2 within the previous set can be determined by the comparison of:

K(0) (K)	CO_2 (ppm)	TOth	Req Blkt (W/m²)	Req-Calc (W/m²)
288.350	632	0.21401	152.979	0.0010
288.834	1,264	0.21401	155.617	-0.0014
288.834	632	0.21401	153.990*	1.6193

* 153.990 is the actual Blanket for this non-equilibrium state.

The third row minus the first row shows the increase of 1.01 W/m² in Blkt(0) from the temperature increase alone, without any increase in CO_2. The discrepancy of 1.62 W/m² (62%, or 1% of the total) is definitively attributable to the CO_2 increase.

Set#2: Only CO_2, with decreases in TOth

For this set, TOth was arbitrarily decreased by 3.75% whenever CO_2 doubled and by 5% whenever CO_2 decreased by 2.5 times. The goal was not to mimic reality but to see what the result would be if TOth decreased symmetrically with $LOG(CO_2)$ increases. For the early steps the $\Delta K(0)$ double is itself double the previous doubling (in Set#1) but the chosen percentage decreases in TOth should not have been directly linear but should have been linear in relation to TOth's approach to its asymptote of 0.0.

K(0) (K)	CO_2 (ppm)	TOth	Req Blkt (W/m²)	Req-Calc (W/m²)
287.642	252.8	0.21756	149.143	-0.0007
288.0	380	0.21401	151.079	0
288.711	632	0.2067	154.946	0.0005

$\Delta K(0) = 1.069 \rightarrow 0.81K$ per doubling
Average $\Delta K(0)$ per 100 ppm = 0.28K

289.564	1,264	0.1989	159.622	-0.0014

$\Delta K(0) = 0.853 \rightarrow 0.85K$ per dblng $\Delta Abs = 2 \cdot \Delta Blkt(0) = 9.35$ W/m²

290.469	2,528	0.1915	164.629	-0.0011

$\Delta K(0) = 0.905 \rightarrow 0.91K$ per doubling

291.782	6,320	0.1819	171.977	-0.0004

$\Delta K(0) = 1.313 \rightarrow 0.99K$ per dblng

K(0) increases from 0.80K per doubling for the lower ppm to 0.99K for the higher ppm for this aggressive reduction in TOth. Overall, $\Delta K(0)$ has increased by 4.1K for a 25X increase in CO_2 (as compared to 2.5K in the first set).

Set#3: CO_2, with introduction of water vapor (TPW= 2cm) and convection

The challenge is to determine what the quantitative effects of H_2O are. The text in the main body of the document indicates that H_2O concentrations in the atmosphere have not been increasing over the past 60 years nor have average annual rainfalls. "Record" events cannot be surmised to be upsetting the average state.

In this set the number of columns has been increased (and the number of rows has been increased).

K(0) (K)	CO_2 (ppm)	H_2O (cm)	Conv. (W/m²)	TOth	Req Blkt (W/m²)	Req-Calc (W/m²)

K(0), CO_2, Required Blanket, and Required-Calculated are all the same as before. H_2O represents water vapor below Z' = 0.40 and is a specific absorber, with its assigned value having no scientifically related correlation to the assigned value for CO_2. Convection (Section Two) is the additional blanket contribution from the Tropics; it does not provide absorption inputs, but it contributes to BBR(0) to the same extent as do the blanket contributions from absorptions. TOth represents absorbers other than CO_2 or H_2O, and is given a tentative value of about 0.9. (It has a tentative (adjustable) value just so that the Convection Blanket can be given in integer units.)

This set was also discussed in Section Two of the main text, but more examples are given here. All calculations use the same spreadsheet discussed in **Appendix C**.

CO_2 values of 252.8 and 632 ppm will primarily be used to emphasize the roles of water absorption and blanket. Other values of CO_2 are used to demonstrate the sensitivity of the final results to the value of the convection blanket. The selection of the convection blanket might present a more important condition for establishing equilibrium at low CO_2 concentrations than for higher CO_2 concentrations. Nature might be expected to have higher convection at higher overall temperatures but the "backpressure" of increased local heating in the upper latitudes might reduce the net convection from the Tropics. (With a diminished ease of outlet, a restrained convection might cause surface temperatures in the Tropics to decrease less than otherwise would be the case for a stronger convection. This is an observation that should be looked for.)

A discussion of the Poles has been intentionally disregarded, but for most non-scientists it is their only "proof" of disastrous global warming. It is constantly claimed that the *minimum extent* of the North Pole's (floating) ice is steadily receding (by about 0.44% a year). Yet its minimum extent over the length of a year has been essentially constant for ten years. Scientists regularly point out that the melting of polar floating ice is correlated with and dominated by the warmth of the ocean currents that pass beneath. The ice, only a few meters thick, is being melted from beneath. Ocean currents are particularly changeable and few would pick a spot in the ocean as an arbiter of global atmospheric warming.

Figure 37 in Section Two definitively shows that temperatures at the North Pole are more than 25C (!!!) higher than those at the South Pole. (Independently of this, average temperatures throughout the northern hemisphere are consistently higher than for the southern hemisphere simply because the southern hemisphere is dominated by water and the larger land mass in the northern hemisphere has a greater effect on surface heat from the sun's input.

Figure 37 shows that the average temperature at 25 Lat North equals the average temperature at 10 Lat South.) In addition to the effects of larger northern land mass, the Arctic sea is nearly surrounded by land mass and so its sea better retains its warmth. Antarctic ice is susceptible to melting by ocean flow only along the shelf (of floating ice and small islands) that reaches up towards South America. This ice finger reaches a Latitude that approximates the southern end of Newfoundland in the north and it is proximate to a large expanse of southern ocean. There are, however, no serious discussions of meaningful melting and reduction of the Antarctic's *continental* buttress of mile-deep thicknesses of ice and snow. Antarctic ice, with 90% of the earth's ice, is actually increasing slowly with time.

Even the reality of some increase in North Polar zone temperature over a long period cannot be said to apply to the entire globe. The North Pole zone is particularly sensitive to the positive feedback of the albedo effect. Once any previously-reflecting snow and ice is melted, more sunlight-absorbing water is exposed to the sun. The temperature rises, more ice is melted, etc. Any flattening of this feedback growth worldwide can be an indicator of the possible end of a warming Age and the beginning of a new Ice Age. A more mundane possibility is that any current increase in melting may not be associated with any extension of the warming age (*or, to any meaningful extent, by any increase in CO_2*), but by an increase in "carbon blackening" of the snow and ice cover. Carbon particulates are pollutants in more ways than one, but CO_2 is, fundamentally, not a pollutant any more than the more dominant water vapor is a pollutant.

The first rows of data for this third set are all for an atmospheric water content of 2 centimeters. This is close to the global average — to the extent an average has meaning for water vapor. The average rainfall per day is only 2 to 3 mm per day (0.2 cm to 0.3 cm per day). Rainfall days of 1.0 to 3.0 cm separated by 5 to 11 days of water vapor increases can easily match the 2 cm overall atmospheric average for water vapor and the rainfall average of 0.2 to 0.3 centimeters per day. Stull et al provide absorption data for both 2 cm and 5 cm water content and so the effects of 5 cm of water will also be examined.

Any increase in reflection of incoming sunlight with higher water content (in the condensed levels of the atmosphere), which would reduce the 239 W/m^2 for the sun's net input, is not considered.

Global Warming Temperatures and Projections

K(0) (K)	CO_2 (ppm)	H_2O (cm)	Conv. (W/m²)	TOth	Req Blkt (W/m²)	Req-Calc (W/m²)
287.642	252.8	0	0	0.21756	149.143	-0.0007
287.642	252.8	2	0	0.21756	149.143	-7.3974
287.642	252.8	2	-7.3974	0.21756	149.143	0.0000

Note the direct relationship between ΔConvectionBlanket and ΔReq-Calc

287.642	252.8	2	0	0.88125	149.413	80.0001
287.642	252.8	2	80	0.88125	69.143	0.0001
287.642	632	2	80	0.88125	69.143	-5.4826

The Req-Calc conclusion of -5.5674 W/m² doesn't say much except that the introduction of 632 ppm increases absorption by 2 • 5.57 = 11.14 W/m².

| 288.9414 | 632 | 2 | 80 | 0.88125 | 76.205 | 0.0000 |

The actual increase in K(0) is 1.30K. Absorption increases by
2 • ΔBlkt = 2 • 7.06 = 14.12 W/m². 14.12/10.9 = 1.295K.
ΔK(0) is 0.99K per doubling of CO_2.
Average ΔK(0) per 100 ppm of CO_2 = 0.34K.

Extracting information from the tabular data

More information can be obtained from the previous data. What follows initially is a tutorial with assumed data. The Tcomb components employ the product of T values at a constant 300K and without adiabatic effects. This Tcomb and its *eff*T components are also employed in the calibration plots. The *actual* transmitted and absorbed powers are, however, significantly lower than what these 300K components might seem to predict.

Assume, for the tutorial: *eff*TMol1 = 0.20, *eff*TMol2 = 0.7, and Tbkgnd = 0.857. Tbkgnd might be a product of other molecules, but allow their T product to be 0.857. *Eff*TConv will have its own value, but, for the moment, it is incorporated within *eff*TMol1.

Tcomb = 0.12 and (1-Tcomb) = 0.88. This Tcomb is compatible with K(0) = 291.08K (see **Figure 16**), P(0) = 407 W/m², Blkt(0) = 168 W/m². Allow the Convection contribution to be 80 W/m²; the absorption contribution to Blkt(0) is 88 W/m².

Blanket power associated with Mol1 and Convection might appear to be ≅ ((1-0.2)/(1-0.12)) • 168 = 152.73. Blanket power associated with Mol2 ≅ ((1-0.7)/0.88) • 168 = 57.27. Blanket power associated with Bkgnd is ≅ ((1-0.857)/0.88) • 168 = 27.3. These results are NOT correct because they represent Blanket values that would be obtained if each were acting as solo absorbers (at constant temperature). Their sum is 237.3 W/m², and a suggested common correction factor is 168/237.3 = 0.708. This might seem to be a large correction factor, but (1-Tcomb) is already 0.88 and the asymptote for 100% absorption is 1.0.

So if we go back to the previous numbers we come up with new numbers. The Blanket associated with Mol1 goes from 152.73 to (108.13-80) = 28.13. Convection is 80. The blanket for Mol2 goes from 57.27 to 40.55, and that for Bkgnd goes from 27.3 to 19.33. The new total sum is 168 and the absorption sum is 88, as they should be.

Let us return to the actual calculations preceding the tutorial and determine some approximate values for the *eff*ectiveT components. **The original conditions (from direct simulation) are:**

{CO_2 = 632;   water = 2cm;   K(0) = 288.9414;   P(0) = 395.205;
         Blkt(0) = 156.205; Conv = 80; AbsBlkt(0) = 76.205}
(1-Tcomb) for K(0) = 288.9414 is 0.855 → Tcomb = 0.145.
Tcomb = (*eff*TCO_2 for 632 ppm) • (*eff*TH_2O+Conv for
    2 cm) • (Tbkgnd)
*eff*TCO_2 for CO_2 = 632 ppm ≅ 0.7244; *eff*TH_2O unknown
*eff*Tbkgnd for the remaining background is 0.881 (an estimate)
Therefore, (*eff*TH_2O+Conv for H_2O = 2 cm) = 0.145/
    (0.7244 • 0.881) = 0.227.

Absorption by H_2O+Convection associated with Blkt power might be expected to be ((1-0.227)/0.855) • 156.205 = 141.22.

Absorption by CO_2 might be expected to be ((1-0.7244)/0.855) • 156.205 = 50.35.

Global Warming Temperatures and Projections

Absorption by Background might be expected to be
$((1-0.881)/0.855) \cdot 156.205 = 21.74$.

The sum equals 213.31, but the actual sum is 156.205 W/m². The correction factor is 0.732.

Absorption by H_2O plus Convection $\cong 103.37$ W/m².
Absorption by $H_2O \cong 23.37$.
Absorption by $CO_2 \cong 36.86$.
Absorption by Background $\cong 15.9$.
The sum for the three absorptions equals 76.1 (as it should).

(Abs by H_2O)/(Abs by remaining background) = 23.37/15.9 = 1.45
(Abs by CO_2)/(Abs by remaining background) = 36.86/15.9 = 2.32
(Abs by H_2O)/(Abs by all else) = 23.37/(36.86+15.9) = 0.44
(Abs by CO_2)/(Abs by all else) = 36.86/(23.37+15.9) = 0.94 $\cong$ 1
(Abs by $H_2O + H_2O$-created Convection)/(Abs by all else) = (23.37+80)/(36.86+15.9) = 1.96 $\cong$ 2
(Abs by CO_2 of 632 ppm)/(Abs by H_2O of 2 cm total up to a height of Z' = 0.4) $\cong$ 36.86/23.37 = 1.58 $\cong$ 1.5

For this example, CO_2 absorption is approximately 50% more than H_2O absorption. When convection produced by the condensation of water vapor in the upper altitudes is included with water vapor absorption, then water dominates by a factor of 2 over all other absorbers. CO_2 absorption represents 25% of the total Blanket, but nearly all of that absorption occurred at very low ppm of CO_2. A full doubling of the present levels of CO_2 will cause less than an additional 2% increase to the blanket. 5 cm of H_2O (next section) adds only another 3.53 W/m² of absorption, which includes the *effect* of a temperature rise of about 0.6K, which increases all absorptions. {Caveat: this summary is not definitive, since assumptions are made throughout the calculations.} A simple summary would say that Convection is providing 50% of the blanket, with 50% provided by CO_2, H_2O and background in that order.

Set#4: CO_2, with water vapor comparisons (TPW = 5 cm vs. 2 cm)

The next set of data considers average precipitable water vapor (PWV) levels of 5 cm. This would never be a true average state, but the calculations are valuable in assessing the sensitivity of K(0) and Blkt(0) to PWV.

K(0) (K)	CO_2 (ppm)	H_2O (cm)	Conv. (W/m²)	TOth	Req Blkt (W/m²)	Req-Calc (W/m²)
287.642	252.8	0	0	0.21756	149.143	-0.0007
287.624	252.8	2	82	0.8979	67.046	-0.0009
288.228	252.8	5	82	0.9	70.316	0.008
288.896 (+1.27)	632	2	82	0.9	73.958 (+6.91)	0.0000
289.5577 (+1.33)	632	5	82	0.9	77.587 (+7.27)	0.0004
290.101 (+1.21)	1,264	2	82	0.9	80.587 (+6.63)	-0.0003
290.777 (+1.22)	1,264	5	82	0.9	84.344 (+6.76)	-0.0007
291.555 (+1.45)	2,528	2	82	0.9	88.699 (+8.11)	-0.0009
292.248 (+1.47)	2,528	5	82	0.9	92.607 (+8.26)	-0.0007

The ΔK(0) per doubling of CO_2 for each successive 2 cm pair = {0.96, 1.15, 1.45}.
The ΔK(0) per doubling of CO_2 for each successive 5 cm pair = {1.01, 1.22, 1.47}.

Note how the ΔK(0) for doubling are barely perturbed by the 2.5X increase in H_2O.

Set#5: Authenticity of temperature measurements and effects of convection values

It can rationally be argued that any reduction in Convection has the same effect as twice that reduction in Absorption since either has the effect of reducing Blkt(0).

This will be trivially demonstrated below. What cannot be proven here, however, is whether an increase of absorption because of either CO_2 or H_2O will either automatically increase or decrease the Convection value.

One logical supposition is that an increase in CO_2 or H_2O will obviously tend to increase temperature locally, that the increase should apply to the Tropics also, that the Tropics' "crown" will increase and its convection flow towards the Poles will increase. An increased flow will increase Convection contributions at all Latitudes. Everything might be seen as working towards increasing temperatures (not a "runaway," just an increase).

However, it is a built-in temperature difference that increases a convection flow. Higher Latitudes have their own evaporations that increase and that transfer heat to the upper atmosphere. The convection flow from the Tropics might be reduced, the "crown" would want to increase, but more rainfall in the Tropics would both reduce the net convection from the Tropics and raise the temperature at the surface of the Tropics and in the Tropical oceans. Are any scientifically verified results of annual Tropics temperature changes relative to higher Latitude changes being uncovered?

This author believes there is insufficient authenticity in the officially stated worldwide averages in temperature. There is an increasing variety of temperature-measuring devices, an increasing attention paid to "corrections" and interpretations of earlier data, and what appears to be an overwhelming difficulty in transforming ocean samples into atmospheric temperatures, and in incorporating these new measurements with existing data packages. Yet final results in temperatures or temperature variations are stated, with apparently great confidence, in hundredths of a degree Centigrade. Who is verifying that proper weighting is being given to the counts *of stations in all of the geographical area divisions of the globe?* Are all temperatures measured in areas with high human generation of local heat discarded, and if not, why not? Specifically, is there a temperature index that employs samples only in the lowly populated (and unpopulated) areas? *Are there any temperature indices that have the continuity of looking at the same locations with the identically same measurement standards for the past 50 to 100 years?*

Since oceans are the great equalizers, is there any progress in establishing scheduled Latitude and Longitude scans of all the oceans by means of a single standardized technique, without relying so much on individual ships, with different equipment, on their normal routes to supply the "ocean" data? It would seem better to have less data with more reliability than more data. "Earthshine" ("earthlight" reflected from the earth) onto the supposedly "dark" areas of the moon is another well-established technique. The corresponding "dark moonshine" (second reflection, not the sunlight reflection from the bright portion of the moon) from the moon can be accurately measured with earth equipment. With the corresponding knowledge of the area of the earth providing this reflection, and the known intensity of the sun onto the earth, the average albedo of that entire area of the earth can be calculated; since the earth is rotating and the area providing the earthshine is changing, more accurate calculations of albedo per smaller unit area can be determined in much the same manner as a medical CAT scan. The uniform topography of the North Pole provides a perfect area for examination. It has already been established that small changes in albedo have a much greater effect on local $K(0)$ than do large changes in CO_2. Blaming Pole melting on CO_2 is seriously affecting the ability of the populace to understand the true issues.

And what is the actual definition of "average" temperature, and has the definition been changing? "Maximum" temperatures are constantly advertised, but not Minimum temperatures or Maximum Average Temperatures that encompass an entire 24 hour day. It is a scientific fact that even with a mathematical normal distribution *that is perfectly identical over time*, the introduction of more and more data samples will always produce new "record" values for the outermost points of the actual observations. The entire area of

Global Warming Temperatures and Projections

temperature measurement needs a MacArthur Prize level of examination of its scientific merit, its accuracy, and the integrity of its "averaging."

In the following set the Local Evaporation (with total return) is changed from 0 to 2.5 for CO_2 = 632 ppm and H_2O = 5 cm and from 0 to 4 for CO_2 = 1,264 and H_2O = 5 cm The accompanying Conv values are {82, 79.5} for CO_2 = 632 and {82, 78} for CO_2 = 1,264. The sum of Local condensation and return PLUS Tropics Convection is always 82. There are no claims that these assignments are correct, but only a plausibility claim that Tropics convection need not necessarily increase when K(0)'s increase because of local water evaporation.

K(0) (K)	CO₂ (ppm)	H₂O (cm)	Loc evap & return (W/m²)	Tropic Conv (W/m²)	TOth	Req Blk
289.5577	632	5	0	82	0.9	77.587
290.777	1,264	5	0	82	0.9	84.344
289.5577	632	5	2.5	79.5	0.9	77.587
290.777	1,264	5	4	78	0.9	84.344

This result was foreordained. The Tropics Convection contribution to Blkt(0) may decrease as CO_2 and K(0) values increase as long as local condensation in the atmosphere counters the convection flow. In effect, the local condensation contributes to the Convection bus and the same net Convection contribution may be made to Blkt(0).

Alternative co-scaling of LOG2(CO₂ ratio) and (1-Tbkgnd)

The previous changes in Tbkgnd (= TOth) from the base line value of 0.21401 for the baseline of {K(0) = 288.0K, CO_2 = 380 ppm,

Tbkgnd = 0.21401} for the lower and upper CO_2 values of 252.8 and 632 ppm were adjustments that doubled the $\Delta K(0)$ slope from 252.8 ppm to 632 ppm from the slope when Tbkgnd was kept fixed at 0.21401.

Since higher values of CO_2 always imply higher absorptions it is more appropriate to scale (1-Tbkgnd) rather than Tbkgnd since there is an asymptote of 1.0 for (1-Tcomb). The appropriate values that increase $\Delta K(0)$ per doubling of CO_2 from 0.40K to 0.80K are:

K(0) (K)	CO_2 (ppm)	H_2O (cm)	Conv (W/m²)	Tbkgnd
287.530	252.8	0	0	0.21979
288.0	380	0	0	0.21401
288.587	632	0	0	0.20920

The ratio of (1-Tbkgnd) values for a CO_2 ratio of 2.5 is 1.0136. That ratio is 1.01088 for a CO_2 ratio of 2.0.

Figure AE1 gives Power and K plots when CO_2 = 252.8 ppm. **Figure AE2** gives Power and K plots when CO_2 = 632 ppm. No direct calculations for H_2O vapor absorption or for Tropics convection are included; they are enveloped within Tbkgnd.

The following predictive table can be set up:

CO_2	CO_2 ratio	Tbkgnd	(1-Tbkgnd)	(1-Tbkgnd) ratio	Tbkgnd ratio
252.8		0.21979	0.78021		
632	2.5	0.20920	0.7908	1.0136	0.9518
1,264	2.0	0.2006	0.7994	1.01088	0.9589
2,528	2.0	0.1919	0.8081	1.01088	0.9566

Global Warming Temperatures and Projections

CO_2	CO_2 ratio	Tbkgnd	(1-Tbkgnd)	(1-Tbkgnd) ratio	Tbkgnd ratio
6,320	2.5	0.1809	0.8191	**1.0136**	0.9427
12,640	2.0	0.1720	0.8280	**1.01088**	0.9404

Simulations will provide K(0) for each of these cases. Each K(0) is related to a specific (1-Tcomb). *Eff*TCO$_2$ = [1-(1-Tcomb)]/Tbkgnd = Tcomb/Tbkgnd. These *eff*TCO$_2$ values can then be employed when later simulations include H$_2$O vapor and additional Blanket produced by convection from the Tropics.

The results of these simulations produce the following table. (The (1-Tbkgnd) ratio has an assumed correlation to the CO$_2$ ratio.)

CO_2 (ppm)	K(0)	ΔK(0)	ΔK(0) per doubling	Tbkgnd	(1-Tbkgnd) ratio	(1-Tcomb)	*eff*TCO$_2$
252.8	**287.530**			0.21979		**0.83050**	0.7712
632	**288.587**	1.057	0.80	0.20920	1.0136	**0.84845**	0.7244
1,264	**289.4813**	0.8943	0.894	0.20060	1.01088	**0.85944**	0.7007
2,528	**290.4502**	0.9689	0.969	0.19190	1.01088	**0.87292**	0.6622
6,320	**291.8277**	1.3775	1.043	0.18091	1.0136	**0.89148**	0.6000
12,640	**292.9423**	1.1146	1.115	0.17200	1.01088	**0.90563**	0.5487

Note how difficult it is for TCO$_2$ to reach 0.50. The total temperature increase for a 50X (5,000%) excursion in CO$_2$ is 5.4C. With three more doublings, and a projected increase of another 4C to 5C, 50% of the earth's oxygen (100,000 ppm) would have been converted to CO$_2$.

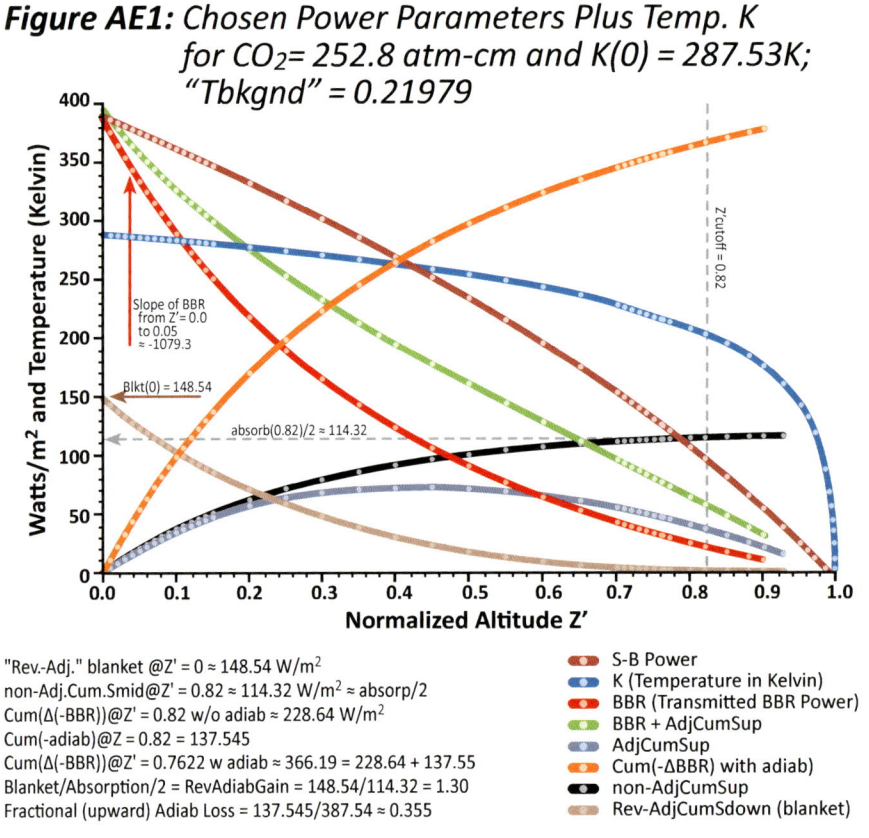

Figure AE1: Chosen Power Parameters Plus Temp. K for CO_2 = 252.8 atm-cm and K(0) = 287.53K; "Tbkgnd" = 0.21979

Graph annotations:
- Slope of BBR from Z' = 0.0 to 0.05 ≈ -1079.3
- Blkt(0) = 148.54
- absorb(0.82)/2 ≈ 114.32
- Z' cutoff = 0.82
- X-axis: Normalized Altitude Z'
- Y-axis: Watts/m² and Temperature (Kelvin)

"Rev.-Adj." blanket @Z' = 0 ≈ 148.54 W/m²
non-Adj.Cum.Smid@Z' = 0.82 ≈ 114.32 W/m² ≈ absorp/2
Cum(Δ(-BBR))@Z' = 0.82 w/o adiab ≈ 228.64 W/m²
Cum(-adiab)@Z = 0.82 = 137.545
Cum(Δ(-BBR))@Z' = 0.7622 w adiab ≈ 366.19 = 228.64 + 137.55
Blanket/Absorption/2 = RevAdiabGain = 148.54/114.32 = 1.30
Fractional (upward) Adiab Loss = 137.545/387.54 ≈ 0.355

Legend:
- S-B Power
- K (Temperature in Kelvin)
- BBR (Transmitted BBR Power)
- BBR + AdjCumSup
- AdjCumSup
- Cum(-ΔBBR) with adiab
- non-AdjCumSup
- Rev-AdjCumSdown (blanket)

Multiple possibilities exist for future calculations. Only one example is given within this document:

K(0)	(1-Tcomb)	CO_2	effTCO₂	H_2O (cm)	Tbkgnd	Add'l conv. Blkt	eff(TH₂O *TConv)
288.587	0.8465	632	0.7337	5	0.87462	75	0.2392

Figure AE3 gives the Power and K plots. Note that Tcomb for **Figure AE2** = 0.7337 • 0.20920 = 0.1535. Tcomb for **Figure AE3**

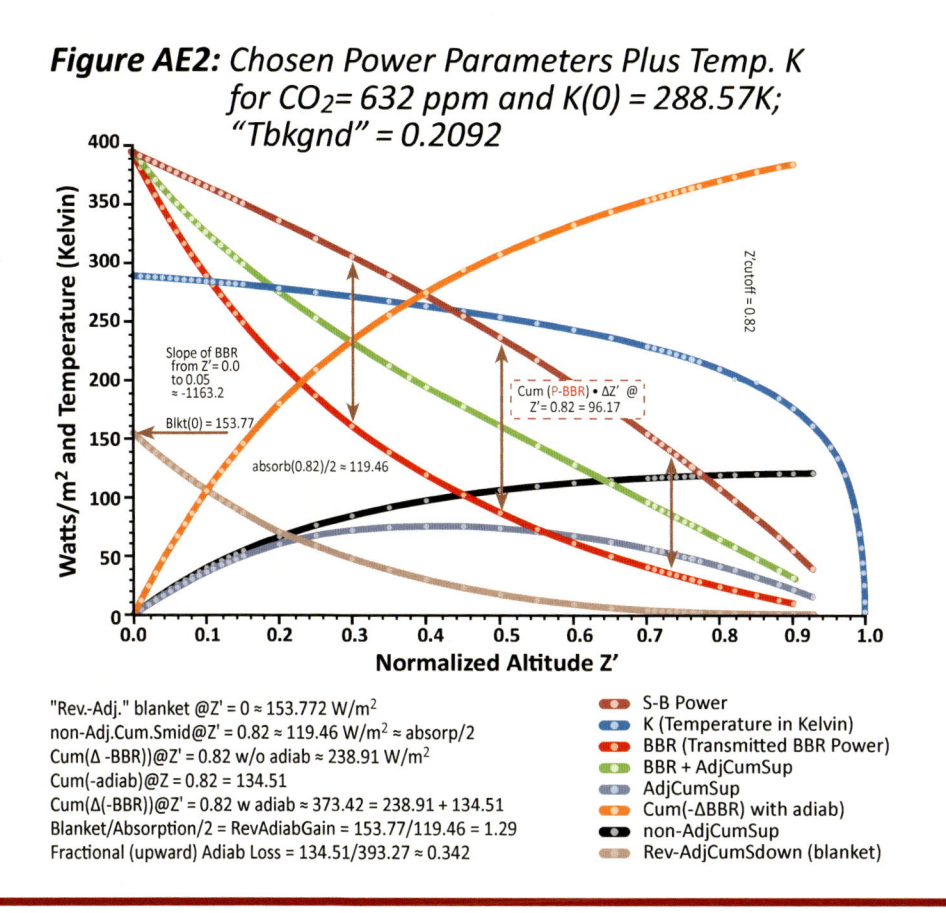

Figure AE2: Chosen Power Parameters Plus Temp. K for CO_2= 632 ppm and K(0) = 288.57K; "Tbkgnd" = 0.2092

"Rev.-Adj." blanket @Z' = 0 ≈ 153.772 W/m²
non-Adj.Cum.Smid@Z' = 0.82 ≈ 119.46 W/m² ≈ absorp/2
Cum(Δ -BBR))@Z' = 0.82 w/o adiab ≈ 238.91 W/m²
Cum(-adiab)@Z = 0.82 = 134.51
Cum(Δ(-BBR))@Z' = 0.82 w adiab ≈ 373.42 = 238.91 + 134.51
Blanket/Absorption/2 = RevAdiabGain = 153.77/119.46 = 1.29
Fractional (upward) Adiab Loss = 134.51/393.27 ≈ 0.342

- S-B Power
- K (Temperature in Kelvin)
- BBR (Transmitted BBR Power)
- BBR + AdjCumSup
- AdjCumSup
- Cum(-ΔBBR) with adiab)
- non-AdjCumSup
- Rev-AdjCumSdown (blanket)

(which has the same K(0)) is 0.7337 • 0.2392 • 0.87462 = 0.1535.

The water contribution (both absorption and convection) is (1-0.2392) = 0.7608 and is three times more than the CO_2 contribution (1-.07337) = 0.2663. But it is important to realize that this CO_2 contribution is only 0.0375 (16%) more that its contribution already was when CO_2 was 2.5 times lower at 252.8 ppm. **From this point of view it can be said that the overall water contribution,** (which, of course, is constantly being recycled as "new" water) **is twenty times more than the *additional* CO_2 contribution for this range.** This truth cannot be ignored.

The great variability in water vapor over locales and over time

Appendix E

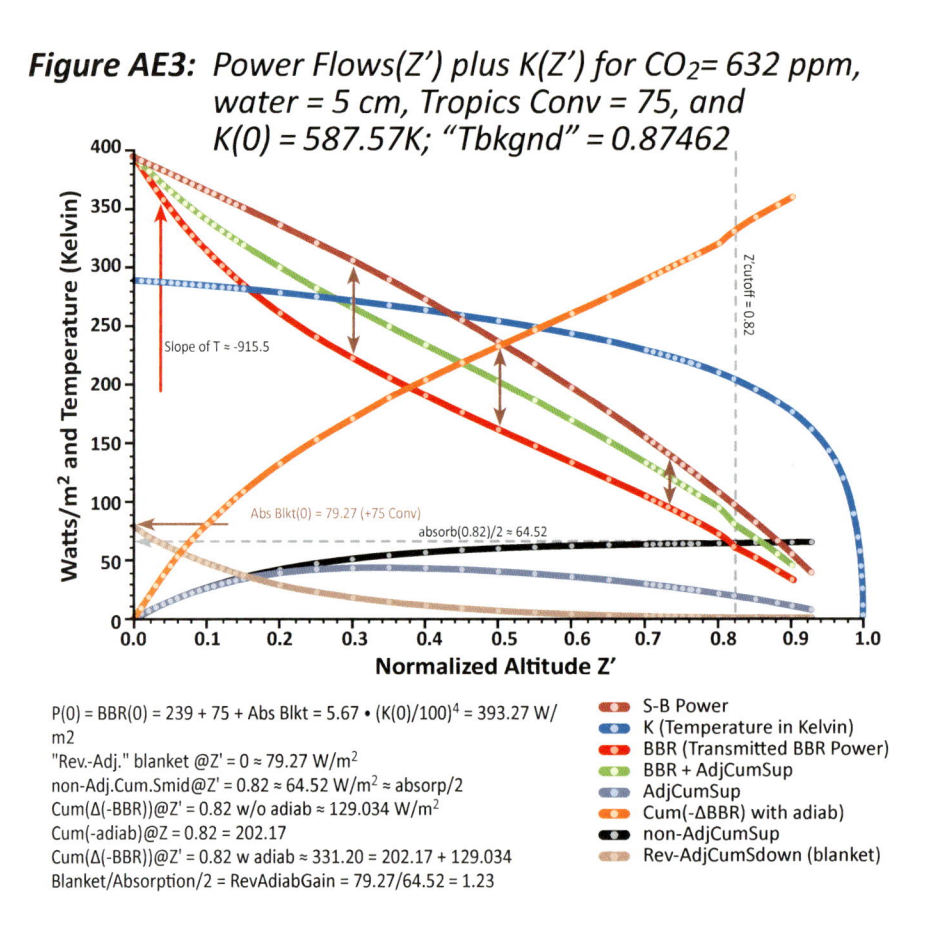

Figure AE3: Power Flows(Z') plus K(Z') for CO_2 = 632 ppm, water = 5 cm, Tropics Conv = 75, and K(0) = 587.57K; "Tbkgnd" = 0.87462

P(0) = BBR(0) = 239 + 75 + Abs Blkt = 5.67 • (K(0)/100)⁴ = 393.27 W/m²
"Rev.-Adj." blanket @Z' = 0 ≈ 79.27 W/m²
non-Adj.Cum.Smid@Z' = 0.82 ≈ 64.52 W/m² ≈ absorp/2
Cum(Δ(-BBR))@Z' = 0.82 w/o adiab ≈ 129.034 W/m²
Cum(-adiab)@Z = 0.82 = 202.17
Cum(Δ(-BBR))@Z' = 0.82 w adiab ≈ 331.20 = 202.17 + 129.034
Blanket/Absorption/2 = RevAdiabGain = 79.27/64.52 = 1.23

- S-B Power
- K (Temperature in Kelvin)
- BBR (Transmitted BBR Power)
- BBR + AdjCumSup
- AdjCumSup
- Cum(-ΔBBR) with adiab
- non-AdjCumSup
- Rev-AdjCumSdown (blanket)

also show how difficult it is to associate so much assumed importance to CO_2 itself.

*Eff*TH_2O for H_2O = 5cm (which is higher than an average level of water) can be extracted in another calculation with the convection Blanket being ignored (except for its folding into a new Tbkgnd that includes TConv). Such a calculation yields: TConv • Tbkgnd = 0.27025 and *eff*TH_2O (from Absorption) = 0.7741. With the knowledge that *eff*TH_2O • TConv = 0.2392, TConv for Conv = 75 equals 0.309.

Tcomb for K(0)= 288.587K is 0.1535. Therefore,

$eff\text{TCO}_2 \cdot eff\text{TH}_2\text{O} \cdot \text{TConv} \cdot \text{Tbkgnd} = 0.1535.$

Since $0.7337 \cdot 0.7741 \cdot 0.309 \cdot 0.87462 = 0.1535$, it would appear that everything was carried out correctly.

*Eff*TH$_2$O is higher (less absorption) than expected; this may be the result of the author's assumption of including the possibility of water vapor absorption occurring only at altitudes below $Z' = 0.4$. Absorption by water droplets is not considered; yet the value of 5 cm is an integration of water over all altitudes. None of these results can be considered to be correct in any absolute sense. They simply point to a means for better calculations if input data is more sophisticated.

It also needs to be restated that no definable connections were found in this study between CO$_2$ increases and the associated small increases in temperature with any definable magnitude of possible effects on water and, somehow, further temperature increases. The models developed here for CO$_2$ had their direct effects arbitrarily multiplied by a factor of two, and still no calamitous effects are observable. Water's effects on weather (climate change) are dramatic and extensive, primarily because of the energy exchanges when water is evaporated and vapor is condensed. Nothing like that is possible with CO$_2$. If a pollutant is defined by its potentially harmful effects then water and wind are much more damaging than CO$_2$. The convection bus heat at higher altitude heat tends to smooth temperatures over Latitude, much as ocean currents also act to transfer heat. There is much more to be learned about the vagaries of weather, but there is little likelihood that energy applications or energy reductions by mankind can ever produce any control over the weather.

Quick references (5 tables)

Possible ways to increase K(0) by 0.1°C [or 1.0°C]

Sun's input increase	240 W/m² → 240.5 (0.2%) [240 → 245 (2%)]
Earth's core input₂	1.0 W/m² → 1.5 [1.0 → 6]
Overall albedo decrease (from polar carbon increase, reduced ice/snow area or increased forest area)	0.3 → 0.299 (-0.33%) [0.3 → 0.286 (-4.7%)]
Increase of atmos CO_2	380 → ~407 (7.1%) [380 → 760 (100%)]
Increase of atmos H_2O	Similar sensitivities as CO_2
Increase of rain in the Tropics producing an increase of average Blanket in Temperate zones	70 W/m² → 70.5 (0.71%) [70 → 75 (7.1%)]
Comparison comment	Solar and rain percent increases or albedo increases are more effective in increasing K(0) than CO_2 or H_2O vapor increases

Atmosphere and ocean CO_2 levels

1.0 GT (metric gigatons) CO_2	*1.0E15 g, 1.0 Pg (Petagram), 0.27 Pg of C*
GT per ppm of CO_2	*7.9 GT, 2.2 Pg C*
GT of CO_2 in the atmosphere	*3,100 GT, 390 ppm, 3,100 Pg (850 Pg of C)*
Total weight of the atmosphere	*5,300,000 GT, 5,300,000 Pg*
Increase in atmos CO_2 per year	*1.0 to max of 2 ppm; 8 to 16 GT; 0.25% to 0.5%*
200 year projection	*600 to max of 1,000 ppm*
Average ocean removal per year	*4-5 GT*
Exchange rate	*About 10% of atmos CO_2 is regularly being exchanged molecularly with oceans and land, with ocean release occurring in the Tropics*
Ocean CO_2/atmospheric CO_2	*A ratio of about 50*

Atmospheric water issues

PW or TPW (total Precipitable Water) & [PWV (includes only water vapor)]	*Integral of all H_2O molecules, and their thickness (mm or cm) if they were all to be collected at the surface [PWV considers only water vapor]*
CO_2 and H_2O (vapor) similarities and differences	*Both have absorption windows (and notching), and equivalent absorption sensitivities; CO_2 is stable spatially and temporally. H_2O levels vary by orders of magnitude, and are affected by wind, temperature and sun. The sum of two separate absorptions (in W/m^2) is greater than a parallel (simultaneous) absorption.*
1.0 cm PWV	*About 1,000 ppm_{mass} and about 1,600 ppm_{vol}*
Average PWV's (mm)	*1-8 in the Temperate, 10-25 in the Tropics*
Average rainfalls per day (mm/day)	*About 3.0 (1-3 in the Temperate, 4.5-6 in the Tropics) Almost 600,000 cu km per year. (For the U.S. the total precipitation is 715 mm/yr, or 1,430 cu miles per year)*
Total average water in the atmosphere	*9,200 GT, 9,200 Pg, 9,200 cu km, 2,200 cu mi → ~1.8 cm*
Total rainfall and the turnover ratio of atmospheric water per year	**About 600,000 GT: turnover of 65 times per year.**
Effect of CO_2 on the rainfall turnover	*None*

Power and energy issues

1.0 Watt (Power)	*1.0 Joules (energy)/sec*
1.0 Joules (energy)	*1.0 W-s = ((1/3,600) • Watt-hour = 0.000278 Wh*
1 KiloWh (KWh)	*3.597 J (and 1.0 J = 0.278 KWh)*
Powers of 10	*10^3 → K; 10^9 → G (Giga); $1E^{12}$ → T (Tera); 10^{15} → P (Peta); 10^{18} → E (Eta)*
Ave Power per year	*Energy/ (364 • 24) → <KW>,… ;<TW>; PW>; <EW>*
Burning 1.0 gram of C	*46.0 KJ; creates 3.67 g of CO_2*
Creating 1 ppm of CO_2 (7.9 GT)	*Burning 2.15 Pg of C*
Retaining 1 ppm of CO_2	*Burning about 4 Pg of C*
Burning one million metric tons (1.0 GT, or 1.0 Pg) of C	*46,000 PJ; creates 3.67 GT of CO_2*
Current burn of about 9.8 GT (9.8 Pg) of C per year	*125,000 TWh (or 0.125 EWh); creates 36 GT of CO_2 (at least 50% goes to the oceans and land)*
Average annual power of C burn	*<14.3 TW> (125,000/(364 • 24))*
Annual CO_2 increase that is retained by the atmosphere	*About 4 to 6 GT each year (about a 0.25% to 0.4% increase in total atmos CO_2);* **1 to 1.5 ppm increase per year**

Nature vs. mankind examples

Sun (total atmos and earth abs)	120,000 TW
Solar power (1% of land, 10% conversion efficiency)	300 TW (20 times 15 TW human need)
Wind power capability	Immense at higher altitude (with tethering!!), but, practically, only about 250 TW
Excess upwell, and convection from the Tropics	probably 14,000 TW, and in play (along with the sun) all the time
Power and energy associated with water collection and heating within a hurricane	600 TW, or 52 EJ per day. Possibly 20,000 GT of rain per day. The potential energy for nature simply to raise this water to a height of 1.5 km is about 35 PWh.
Overall power and energy of a hurricane	Wind power of 1.5 TW. Total energy of 10,000 one-megaton bombs
Rotational energy of the earth	$2.19 \cdot 10^{28}$ J, or 21.9 billion EJ. Rotational energy is the solid "battery" for all weather
Can humans (or CO_2) hope to noticeably affect these weather issues?	Particulates affect water nucleation. Carbon deposits affect the earth's albedo at the Poles.

About the author

William T. Lynch is a Fellow in the Institute of Electrical and Electronic Engineers (IEEE) Technical Society, and has B.S., M.S., and PhD degrees from the University of Notre Dame, Massachusetts Institute of Technology, and the University of Princeton. He has presented and published multiple peer-reviewed papers and has over 60 patents. He wrote a prize-winning undergraduate student paper in 1957 on a model for photovoltaic cells for insertion into larger circuit models, and in the late 50's and early 60's became an expert in nuclear radiation while in the U.S. Navy on a tri-military base. He has maintained and advanced his knowledge on all energy and power topics, including wind energy, since then.

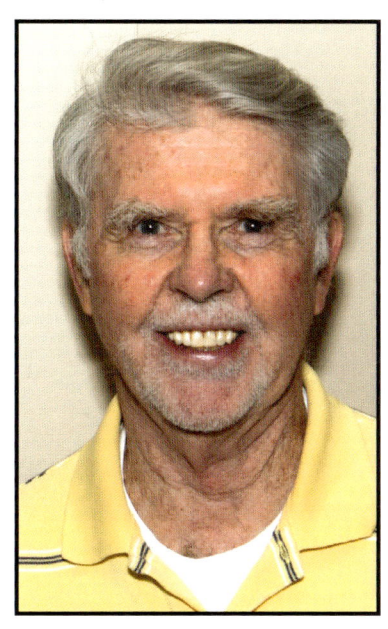

William T. Lynch, PhD

Most of his patents were developed at Bell Telephone Laboratories and were in the areas of solid state devices (transistors),

About the author

integrated circuit designs and their processing implementation, interconnect wiring, optoelectronics devices and their coupling, and unique memory cells and arrays and their access. He became Head of a renowned broad-disciplined Department that addressed all aspects of Very Large Scale Integration (VLSI).

After a first "retirement," he became a Director at the Semiconductor Research Corporation, a non-profit industry consortium that sponsors PhD university research on future needs of the semiconductor industry. He served on national roadmap committees, advised a government committee that oversaw the merit of potential integrated circuit contracts across all government agencies, initiated multiple unique conferences, and oversaw $10 million of university research with the collaborative aid of industry personnel. Some of his most enjoyable creativity has come since his second "retirement." He worked with two tentative start-ups and one actual start-up, resulting in several patent disclosures and five patents.

His special techniques of "normalizing" problems (so that solutions apply to a class of problems) have been applied to traffic, automobile performance, economic portfolios and student testing issues. His major activities have been on education metrics, CO_2 modeling, and climate change, the last two being summarized in this book.

Colophon

This book was created with Adobe InDesign CC. Most of the figures were reproduced as vector drawings in InDesign from charts generated from the author's extensive Excel workbooks. Other figures were supplied as images and their creators are credited.

The body text is 14 point Cambria on 14 point leading. Headings and titles are also set in Cambria.

The book was printed through Lulu.com. For more information about its production or content, contact the author at:
GlobalWarming.Lynch@gmail.com